Topics in Operator Theory

Chicago Lectures in Mathematics

topics in operator theory

Richard Beals

The University of Chicago Press
Chicago and London

Chicago Lectures in Mathematics Series
Irving Kaplansky, *Editor*

The Theory of Sheaves, by Richard G. Swan (1964)
Topics in Ring Theory, by I. N. Herstein (1969)
Fields and Rings, by Irving Kaplansky (1969)
Infinite Abelian Group Theory, by Phillip A. Griffith (1970)
Topics in Operator Theory, by Richard Beals (1971)
Lie Algebras and Locally Compact Groups, by Irving Kaplansky (1971)
Several Complex Variables, by Raghavan Narasimhan (1971)

International Standard Book Number: 0-226-03985-4
Library of Congress Catalog Card Number: 70-147095

The University of Chicago Press, Chicago 60637
The University of Chicago Press, Ltd., London

Published 1971
Printed in the United States of America

To the memory of my father

CONTENTS

PREFACE

These lecture notes developed from a course given at the University of Chicago in the fall quarter of 1969 and the spring quarter of 1970. The aim of the course and the notes was to proceed efficiently from a basic knowledge of bounded linear operators in Hilbert space to some of the deeper and most interesting parts of the theory of linear operators. On the way one glimpses some of the rich interplay between real analysis, function theory, and Hilbert space theory.

The basic knowledge is recapitulated in chapter 1. The functional calculus of unbounded self-adjoint operators is developed directly (without B^*-algebras or reduction to the bounded case) in chapter 2. This is applied in chapter 3 to give the spectral theorem and the theorem of Stone on the generation of unitary groups. The theory of spectral representations and spectral multiplicity of self-adjoint and unitary operators is developed in chapters 4 and 5.

Chapters 6 and 7 deal with discrete and continuous shift operators and some of the surprisingly numerous related questions. Among the latter are the theorem of F. and M. Riesz on analytic measures, the Plancherel theorem, the theorem of von Neumann on representing the canonical commutation relations of quantum mechanics, and the theorems of Beurling and of Lax on invariant subspaces.

Some aspects of the theory of dissipative operators are dealt with in chapters 8 through 11. These include self-adjoint dilations, the functional calculus, and the generation of contraction semi-groups. We proceed directly by analysis of the resolvent operator, rather than by use of the Cayley transform. The theory of the characteristic function, due to Livšic, is treated in chapters 9 and 10. The most interesting special result here is the theorem of Brodskii and Livšic determining all invariant subspaces of the simplest Volterra integral operator.

Some historical and bibliographic notes are appended after chapter 11.

Special thanks are due to Fred Flowers for turning the manuscript into a finished product.

1

BOUNDED OPERATORS IN HILBERT SPACE

1. Notation and definitions, positive operators, projections

All Hilbert spaces considered will be complex, with inner products denoted by (x, y). Then $\|x\| = (x, x)^{1/2}$. The norm and the inner product are related by the Schwarz inequality:

$$(1.1) \qquad |(x, y)| \le \|x\| \, \|y\|$$

and by the polarization identity:

$$(1.2) \qquad 4(x, y) = \|x+y\|^2 - \|x-y\|^2 + i\|x+iy\|^2 - i\|x-iy\|^2 .$$

Note that (1.1) holds even if the form (x, y) is not definite. Also (1.2) holds, with $\|z\|^2$ replaced by (z, z), for any form (x, y) which is linear in the first variable and satisfies $(y, x) = \overline{(x, y)}$.

If $\mathcal{H}$ and $\mathfrak{K}$ are Hilbert spaces, $\mathcal{B}(\mathcal{H}, \mathfrak{K})$ denotes the space of continuous (= bounded) linear operators $S: \mathcal{H} \to \mathfrak{K}$. There are three principal notions of convergence in $\mathcal{B}(\mathcal{H}, \mathfrak{K})$:

uniform: $S_n \Rightarrow S$ if $\|S_n - S\| \to 0$;

strong: $S_n \longrightarrow S$ if $\|S_n x - Sx\| \to 0$, all $x \in \mathcal{H}$;

weak: $S_n \rightharpoonup S$ if $(S_n x - Sx, y) \to 0$, all $x \in \mathcal{H}$, $y \in \mathfrak{K}$.

The adjoint of $S \in \mathcal{B}(\mathcal{H}, \mathfrak{K})$ is the operator $S^* \in \mathcal{B}(\mathfrak{K}, \mathcal{H})$ which is uniquely defined by the equation

$$(1.3) \qquad (x, S^* y) = (Sx, y) , \quad x \in \mathcal{H}, \quad y \in \mathfrak{K} .$$

Then $(\alpha S)^* = \bar{\alpha} S^*$, $(S+T)^* = S^* + T^*$, $(ST)^* = T^* S^*$, $(S^*)^* = S$.

The operator $U \in \mathcal{B}(\mathcal{H}, \mathcal{K})$ is an <u>isometry</u> if $\|Ux\| = \|x\|$, $x \in \mathcal{H}$. In view of (1.2) this is equivalent to $(Ux, Uy) = (x, y)$, all $x, y \in \mathcal{H}$, while by (1.3) it is equivalent to $U^* U = I$, the identity operator in $\mathcal{H}$. If the isometry U is onto, then this implies $U^* = U^{-1}$; such a U is said to be <u>unitary</u>.

The operator $S \in \mathcal{B}(\mathcal{H})$ is <u>self-adjoint</u> if $S^* = S$, or <u>normal</u> if $S^* S = SS^*$. By (1.2) and (1.3), S is self-adjoint if and only if (Sx, x) is real for all x. If $(Sx, x) \geq 0$, all x, S is said to be <u>positive</u>. If S and T are self-adjoint so is $S - T$; if $S - T \geq 0$ we write $S \geq T$.

LEMMA. If $S \geq 0$, then $\|S\| = \sup \{(Sx, x) \mid \|x\| \leq 1\}$.

<u>Proof.</u> Let α be the supremum. Clearly $\alpha \leq \|S\|$. But (1.1) applied to the form (Sx, x) gives $|(Sx, y)|^2 \leq (Sx, x)(Sy, y) \leq \alpha^2$ for $\|x\|, \|y\| \leq 1$. Therefore $\|S\| \leq \alpha$. ∎*

PROPOSITION 1.1. Suppose $\{S_n\}$ is a decreasing (increasing) sequence of self-adjoint operators in $\mathcal{H}$ and $S_n \geq T$ $(S_n \leq T)$, all n. Then there is an $S \geq T$ $(S \leq T)$ such that $S_n \to S$.

<u>Proof.</u> Suppose $S_n \geq S_{n+1} \geq T$, all n. Replacing S_n by $S_n - T$, we may suppose $S_n \geq S_{n+1} \geq 0$ and show that $S_n \to S \geq 0$. Given $x \in \mathcal{H}$, we want to show $\{S_n x\}$ converges. Let $y_{mn} = S_m x - S_n x$ and let

$$(y, z)_{mn} = (S_m y - S_n y, z) \quad , \quad m < n.$$

---

*We shall denote the end of a proof by ∎.

By (1.1) applied to this inner product,

$$(1.4) \qquad \|y_{mn}\|^4 = (x, y_{mn})^2_{mn} \le (x, x)_{mn}(y_{mn}, y_{mn})_{mn} .$$

Now $(x, x)_{mn} = (S_m x, x) - (S_n x, x) \to 0$, since $(S_m x, x)$ is non-increasing and non-negative. Also

$$(y_{mn}, y_{mn})_{mn} = ((S_m - S_n)^3 x, x) \le \|S_m - S_n\|^3 \|x\|^2 .$$

By the lemma, $\|S_m - S_n\| \le \|S_m\| \le \|S_1\|$. Thus $\|y_{mn}\| \to 0$. Let $Sx = \lim S_n x$. Then $(Sx, x) = \lim(S_n x, x) \ge 0$. The case of increasing S_n is reduced to the decreasing case by considering $-S_n$. $\blacksquare$

A <u>projection</u> in $\mathcal{H}$ is an operator P with $P = P^* = P^2$. If $\mathfrak{K} \subseteq \mathcal{H}$ is a closed subspace and $\mathfrak{K}^\perp = \{x \in \mathcal{H} \mid (x, y) = 0, \text{ all } y \in \mathfrak{K}\}$ is the orthogonal complement, then any $x \in \mathcal{H}$ can be written uniquely as $x_1 + x_2$, $x_1 \in \mathfrak{K}$, $x_2 \in \mathfrak{K}^\perp$. The map $x \to x_1$ is the (unique) projection P with $P(\mathcal{H}) = \mathfrak{K}$. Then

$$(x, x) = (x_1, x_1) + (x_2, x_2) \ge (x_1, x_1) = (Px, x) \ge 0,$$

so $0 \le P \le I$. More generally, if $\mathfrak{K}_1, \mathfrak{K}_2$ are closed subspaces with corresponding projections P_1, P_2, then

$$\mathfrak{K}_1 \subset \mathfrak{K}_2 \iff P_1 \le P_2 \iff P_1 P_2 = P_2 P_1 = P_1 .$$

If $\{\mathfrak{K}\}$ is a family of closed subspaces of $\mathcal{H}$, let

$$\wedge \mathfrak{K} = \bigcap \mathfrak{K}, \quad \vee \mathfrak{K} = \text{closed linear span of } (\bigcup \mathfrak{K}).$$

If $\{\mathfrak{K}_n\}$ is an increasing (decreasing) sequence of closed subspaces with projections $\{P_n\}$, then $P_n \to P$ with P the projection on $\vee \mathfrak{K}$ $(\wedge \mathfrak{K})$.

2. Spectrum and spectral radius

An $S \in \mathcal{B}(\mathcal{H})$ is <u>invertible</u> if there is $T \in \mathcal{B}(\mathcal{H})$ with $ST = TS = I$. Then $T = S^{-1}$ is unique. The <u>resolvent set</u> $\rho(S)$ is the set of complex λ such that $\lambda I - S$ is invertible; we shall write $\lambda - S$ for $\lambda I - S$. The <u>resolvent operator</u> is $(\lambda - S)^{-1}$. The <u>spectrum</u> $\sigma(S)$ is the complement of $\rho(S)$ in $\mathbb{C}$. We shall assume here that $\mathcal{H} \neq (0)$.

LEMMA. If $\|S\| < 1$, then $I - S$ is invertible.

<u>Proof.</u> In analogy with the scalar case, consider the series $\sum_{n\geq 0} S^n$. Since $\|S\| < 1$, this converges uniformly. Let $T_n = \sum_{m\leq n} S^m$. Then $(I - S)T_n = T_n(I - S) = 1 - S^{n+1} \Longrightarrow I$. Thus $T_n \Longrightarrow T = S^{-1}$. ∎

PROPOSITION 1.2. Suppose $S \in \mathcal{B}(\mathcal{H})$, $\mathcal{H} \neq (0)$. Then $\sigma(S)$ is non-empty and compact. The map $\lambda \to (\lambda - S)^{-1}$ is holomorphic from $\rho(S)$ to $\mathcal{B}(\mathcal{H})$.

<u>Proof.</u> There are various possible definitions of "holomorphic" (only apparently different); we shall show the existence of a convergent power series expansion. First, if $|\lambda| > \|S\|$, then $\lambda - S = \lambda(I - \lambda^{-1}S)$, which by the lemma is invertible. Thus $\sigma(S)$ is bounded. If $\lambda \in \rho(S)$ and $T = (\lambda - S)^{-1}$, then $\lambda + \varepsilon - S = (\lambda - S)(I + \varepsilon T)$. For $|\varepsilon| < \|T\|^{-1}$, this is invertible and its inverse is given by the convergent power series $\sum(-\varepsilon)^n T^{n+1}$. Therefore $\lambda \to (\lambda - S)^{-1}$ is holomorphic and $\rho(S)$ is open. Finally, suppose $\sigma(S)$ were empty. For each x, y the function

$\lambda \to ((\lambda - S)^{-1}x, y)$ is entire. Moreover for large $|\lambda|$, $(\lambda - S)^{-1} = \lambda^{-1} \sum \lambda^{-n} S^n$ has norm $\leq |\lambda|^{-1}(1 - |\lambda|^{-1}\|S\|)^{-1}$, so this function vanishes at ∞. By Liouville's theorem, $((\lambda - S)^{-1}x, y) \equiv 0$, so $(\lambda - S)^{-1} \equiv 0$, a contradiction. ∎

Let $N(S)$ and $R(S)$ denote the null-space (kernel) and range of S, respectively. It follows from (1.3) that

(1.5) $$R(S)^{\perp} = N(S^*).$$

PROPOSITION 1.3. If S is self-adjoint, $\sigma(S)$ is contained in the real interval $[\alpha, \beta]$, where

$$\alpha = \inf\{(Sx, x) \mid \|x\| = 1\}, \quad \beta = \sup\{(Sx, x) \mid \|x\| = 1\}.$$

If U is unitary, then $\sigma(U)$ is contained in the unit circle $\Gamma = \{\lambda \mid |\lambda| = 1\}$

Proof. Suppose $\lambda \notin [\alpha, \beta]$. Let δ be the distance from λ to the interval. If $\|x\| = 1$,

$$\|\lambda - S\| \geq |((\lambda - S)x, x)| = |\lambda - (Sx, x)| \geq \delta$$

It follows that $\lambda - S$ is 1-1 and has closed range; moreover if $R(\lambda - S) = \mathcal{H}$, then the algebraic inverse is bounded, with norm $\leq \delta$. Now $R(\lambda - S) = N((\lambda - S)^*) = N(\bar{\lambda} - S)$. But $\bar{\lambda} \notin [\alpha, \beta]$, so $\bar{\lambda} - S$ is 1-1, so $R(\lambda - S) = \mathcal{H}$.

If $|\lambda| \geq \|U\| = 1$, then $\lambda - U$ is invertible. If $|\lambda| < 1$, then $\lambda - U = \lambda(1 - \lambda^{-1}U) = \lambda U(U^* - \lambda^{-1})$. Since λU and $U^* - \lambda^{-1}$ are invertible, so is $\lambda - U$. ∎

The spectral radius of S is the radius of the smallest closed disc in $\mathbb{C}$ with center 0 which contains $\sigma(S)$. The following formula for the spectral radius is fundamental.

THEOREM 1.1. For any $S \in \mathcal{B}(\mathcal{H})$,

(1.5) $$\sup \{ |\lambda| \,\big|\, \lambda \in \sigma(S)\} = \lim_{n \to \infty} (\|S^n\|)^{1/n}.$$

Proof. Let $f(n) = \|S^n\|^{1/n}$, and let $r = \inf \{ f(n) \}$. Given $\varepsilon > 0$, take m so that $f(m) < r + \varepsilon$. Expressing an arbitrary n as $pm + q$, $0 \leq q < m$, we get

$$r \leq f(n) \leq f(m)^{pm/n} f(q)^{q/n} < (r+\varepsilon) f(q)^{q/n}.$$

As $n \to \infty$ we get $f(n) < r + 2\varepsilon$. It follows that $f(n) \to r$.

For $|\lambda| > r$, the series $\lambda^{-1} \sum \lambda^{-n} S^n$ converges uniformly to $\lambda^{-1}(1 - \lambda^{-1}S)^{-1} = (\lambda - S)^{-1}$. Conversely, suppose $(\lambda - S)^{-1}$ exists for all $|\lambda| \geq R$. Then $(1 - \lambda S)^{-1}$ exists for $|\lambda| \leq \rho = R^{-1}$. This inverse can be expressed by the Cauchy integral formula; in fact set

(1.6) $$T_\lambda = \frac{1}{2\pi i} \int_{|\mu| = \rho} (\mu - \lambda)^{-1} (1 - \mu S)^{-1} \, d\mu, \qquad |\lambda| < |\mu|.$$

(This integral exists as the Riemann integral of a continuous operator-valued function.) For $x, y \in \mathcal{H}$

$$(T_\lambda x, y) = \frac{1}{2\pi i} \int_{|\mu| = \rho} (\mu - \lambda)^{-1} ((1 - \mu S)^{-1} x, y) \, d\mu$$

$$= ((1 - \lambda S)^{-1} x, y)$$

by the Cauchy integral formula for the holomorphic scalar-valued function $((1 - \lambda S)^{-1} x, y)$. Thus $T_\lambda = (1 - \lambda S)^{-1}$, $|\lambda| < |\mu|$.

On the other hand, $T_\lambda = \sum \lambda^n S^n$ for λ small. The uniqueness theorem applied to $(T_\lambda x, y)$, all x, y, shows that the coefficients are unique. On the other hand, expanding

$(\mu - \lambda)^{-1} = \sum \lambda^n \mu^{-n-1}$ in (1.6) we get a second expansion in powers of λ. Equating coefficients,

$$S^n = \frac{1}{2\pi i} \int_{|\mu| = \rho} \mu^{-n-1}(1 - \mu S)^{-1} d\mu.$$

The obvious estimations gives $\|S^n\| \leq M\rho^{-n} = MR^n$, M independent of n. Thus $r \leq R$. ∎

COROLLARY. If S is normal, then

(1.7) $$\sup\{|\lambda| \mid \lambda \in \sigma(S)\} = \|S\|.$$

Proof. Note first that for any operator T, T^*T is positive, so

$$\|T\|^2 \geq \|T^*T\| = \sup\{(T^*Tx, x) \mid \|x\| = 1\} = \|T\|^2 .$$

In particular, if $T = T^*$ then $\|T^2\| = \|T\|^2$. By induction, $\|T^n\| = \|T\|^n$, all $n = 2^k$. Applying this to S^*S, we get

$$\|S^n\|^2 = \|(S^n)^*S^n\| = \|(S^*S)^n\| = \|S^*S\|^n = \|S\|^{2n} .$$

This and (1.5) give (1.7).

3. Direct sums

Let $\{\mathcal{H}_\alpha\}$ be a family of Hilbert spaces indexed by a set $I = \{\alpha\}$. The direct sum $\oplus \mathcal{H}_\alpha$ is the subset of the Cartesian product set $\prod \mathcal{H}_\alpha = \{x = (x_\alpha) \mid x_\alpha \in \mathcal{H}_\alpha\}$ consisting of those $x = (x_\alpha)$ for which $x_\alpha = 0$ except for at most countably many R, and such that $\sum \|x_\alpha\|^2 < \infty$. This is a Hilbert space with respect to the inner product $(x, y) = \sum (x_\alpha, y_\alpha)$. We shall write the direct sum of two Hilbert spaces $\mathcal{H}$ and $\mathcal{K}$ as $\mathcal{H} \oplus \mathcal{K}$, and the elements as $x \oplus y$, $x \in \mathcal{H}$, $y \in \mathcal{K}$.

If $\mathcal{K} = \oplus \mathcal{H}_\alpha$ and $\mathcal{K}_\alpha = \{x \mid x_\beta = 0, \beta \neq \alpha\}$, then the $\mathcal{K}_\alpha$ are mutually orthogonal and $\vee \mathcal{K}_\alpha = \mathcal{K}$. Conversely, if $\mathcal{K}$ is a Hilbert space and $\{\mathcal{K}_\alpha\}$ is a family of mutually orthogonal subspaces with $\vee \mathcal{K}_\alpha = \mathcal{K}$, then there is a natural unitary map of $\mathcal{K}$ onto $\oplus \mathcal{K}_\alpha$. We shall often simply identify these two spaces.

4. Invariant subspaces

A closed subspace $\mathcal{H}_1$ of a Hilbert space $\mathcal{H}$ is said to be <u>invariant</u> for $S \in \mathcal{B}(\mathcal{H})$ if $S(\mathcal{H}_1) \subset \mathcal{H}_1$. If $\mathcal{H}_1$ is invariant for S, then $\mathcal{H}_1^\perp$ is invariant for S^*. In fact if $x \in \mathcal{H}_1$ and $y \in \mathcal{H}_1^\perp$, then $(S^*y, x) = (y, Sx) = 0$ since $Sx \in \mathcal{H}_1$. The closed subspace $\mathcal{H}_1$ is said to <u>reduce</u> S if both $\mathcal{H}_1$ and $\mathcal{H}_1^\perp$ are invariant for S. This is equivalent to saying that $\mathcal{H}_1$ is invariant for both S and S^*.

Suppose $\mathcal{H}_1$ is a closed subspace of $\mathcal{H}$, $\mathcal{H}_2 = \mathcal{H}_1^\perp$, and P_1, P_2 are the corresponding projections. Then $\mathcal{H}_1$ is invariant for S if and only if $P_1SP_1 = SP_1$. It follows that $\mathcal{H}_1$ reduces S if and only if

$$SP_1 = P_1SP_1 = (P_1S^*P_1)^* = (S^*P_1)^* = P_1S.$$

This chain of equalities holds if and only if $SP_1 = P_1S$ or equivalently $SP_2 = P_2S$.

We say that a closed subspace $\mathcal{H}_1 \subset \mathcal{H}$ is <u>completely non-reducing</u> for $S \in \mathcal{B}(\mathcal{H})$ if it is invariant and contains no non-zero subspace which reduces S. The determination of the invariant subspaces of a given operator S can be reduced to the determination of reducing and completely non-reducing subspaces.

PROPOSITION 1.4. Let $\mathcal{H}_1$ be invariant for $S \in \mathcal{B}(\mathcal{H})$. There are uniquely determined subspaces $\mathcal{H}_2, \mathcal{H}_3 \subset \mathcal{H}$, such that $\mathcal{H}_1 = \mathcal{H}_2 \oplus \mathcal{H}_3$, $\mathcal{H}_2$ reduces S, and $\mathcal{H}_3$ is completely non-reducing for S.

Proof. Suppose $\mathcal{K}_1$ and $\mathcal{K}_2$ reduce S, and let $\mathcal{K} = \mathcal{K}_1 \vee \mathcal{K}_2$. Then $\mathcal{K}$ is clearly invariant, and so is $\mathcal{K}^{\perp} = (\mathcal{K}_1)^{\perp} \cap (\mathcal{K}_2)^{\perp}$. It follows that there is a unique maximal reducing subspace $\mathcal{H}_2 \subset \mathcal{H}_1$, the subspace generated by all the reducing subspaces of $\mathcal{H}_1$. Then clearly $\mathcal{H}_3 = \mathcal{H}_1 \ominus \mathcal{H}_2 = \mathcal{H}_1 \cap (\mathcal{H}_2)^{\perp}$ is invariant and completely non-reducing.

If $\mathcal{H}_1 = \mathcal{H}'_2 \oplus \mathcal{H}'_3$ is another such decomposition, then $\mathcal{H}'_2 \subset \mathcal{H}_2$ by maximality. Now $\mathcal{H}_2 \cap \mathcal{H}'_3$ is invariant, and so is $(\mathcal{H}_2 \cap \mathcal{H}'_3)^{\perp} = \mathcal{H}^{\perp}_2 \vee \mathcal{H}'_2$. Since $\mathcal{H}'_3$ is completely non-reducing, $\mathcal{H}_2 \cap \mathcal{H}'_3 = (0)$. Therefore $\mathcal{H}_2 = \mathcal{H}'_2$, $\mathcal{H}_3 = \mathcal{H}'_3$. ∎

5. The closed graph theorem

Suppose $\mathcal{H}$ and $\mathcal{K}$ are Hilbert spaces (or even Banach spaces) and $S: \mathcal{H} \to \mathcal{K}$ is a linear operator defined on all of $\mathcal{H}$. The <u>graph</u> $G(S) = \{x \oplus Sx \mid x \in \mathcal{H}\}$ is a linear subspace of $\mathcal{H} \oplus \mathcal{K}$. If $S \in \mathcal{B}(\mathcal{H}, \mathcal{K})$ then G(S) is easily seen to be closed. The <u>closed graph theorem</u> states that conversely if G(S) is closed, then S is continuous, i.e. $S \in \mathcal{B}(\mathcal{H}, \mathcal{K})$. Let us sketch the proof.

LEMMA (Baire category theorem). Let X be a complete metric space and $\{X_n\}$ a sequence of closed subsets whose union is X. Then some X_n contains a ball $B(x_o, \varepsilon) = \{x \in X \mid d(x, x_o) \leq \varepsilon\}$.

<u>Proof.</u> Assume the contrary. Inductively, choose $x_n \in X$ and $\varepsilon_n > 0$ such that $\varepsilon_n \to 0$, $B(x_n, \varepsilon_n) \cap X_n = \emptyset$, and $B(x_{n+1}, \varepsilon_{n+1}) \subset B(x_n, \varepsilon_n)$. Let $x_o = \lim x_n$. Then $x_o \notin \bigcup X_n$. ∎

LEMMA. Suppose $S: \mathcal{H} \to \mathcal{k}$ is linear, 1-1, and onto. Suppose S^{-1} is continuous. Then so is S.

<u>Proof.</u> Let X_n be the closure of $\{x \in \mathcal{H} \mid \|Sx\| \leq n\}$. Then $\bigcup X_n = \mathcal{H}$, so some X_n contains a ball $B(x_o, \varepsilon)$. Therefore if $\|x\| \leq 1$ there is a $y \in X_n$ with $\|y - (x_o + \varepsilon x)\| \leq \varepsilon/2$. Let $x_1 = \varepsilon^{-1}(y - x_o)$. Then $\|x - x_1\| \leq 1/2$ and $\|Sx_1\| \leq M = \varepsilon^{-1} \cdot 2n$. Since $\|2(x - x_1)\| \leq 1$ we can choose x_2 with $\|2(x - x_1 - x_2)\| \leq 1/2$ and $\|S(2x_2)\| \leq M$; in other words $\|x - x_1 - x_2\| \leq 1/4$ and $\|Sx_2\| \leq M/2$. Continuing, we get $\{x_n\}$ with $\|x - x_1 - x_2 - \ldots - x_n\| \leq 2^{-n}$ and $\|Sx_n\| \leq 2^{-n-1}M$. Let $y_n = x_1 + \ldots + x_n$. Then $y_n \to x$ and $Sy_n \to z$ with $\|z\| \leq 2M$. Since S^{-1} is continuous, $x = \lim S^{-1}(Sy_n) = S^{-1}z$, $z = Sx$, and $\|Sx\| \leq 2M$. ∎

Now suppose $S: \mathcal{H} \to \mathcal{k}$ is linear and $G(S)$ is closed. As a mapping of $\mathcal{H}$ into $G(S) = \mathcal{k}_1$, S is 1-1 and onto. The inverse mapping is clearly continuous. Therefore S is continuous and so there is a constant c with $\|x \oplus Sx\| \leq c\|x\|$, all $x \in \mathcal{H}$. Since $\|Sx\| \leq \|x \oplus Sx\|$, this shows that $S \in \mathcal{B}(\mathcal{H}, \mathcal{k})$.

2

FUNCTIONAL CALCULUS FOR BOUNDED AND UNBOUNDED SELF-ADJOINT OPERATORS

1. Unbounded linear operators

In general a linear operator from one Hilbert space $\mathcal{H}$ to another $\mathfrak{K}$ is a linear mapping S whose domain D(S) is a linear subspace of $\mathcal{H}$ and whose range R(S) is a linear subspace of $\mathfrak{K}$. If $S, T: \mathcal{H} \to \mathfrak{K}$ are linear and $\alpha \in \mathbb{C}$, then αS and $S+T$ are defined by

$$D(\alpha S) = D(S)\ ,\ (\alpha S)x = \alpha(Sx)\ ,\ x \in D(S);$$

$$D(S+T) = D(S) \cap D(T)\ ,\ (S+T)x = Sx + Tx\ ,\ x \in D(S) \cap D(T).$$

If $S: \mathcal{H} \to \mathcal{H}'$ and $T: \mathcal{H}' \to \mathcal{H}''$, then

$$D(TS) = S^{-1}(D(T))\ ,\ (TS)x = T(Sx)\ ,\ x \in D(TS)\ .$$

If D(S) is dense in $\mathcal{H}$, S is said to be densely defined. It is said to be closed if the graph $G(S) = \{x \oplus Sx \mid x \in D(S)\}$ is a closed subspace of $\mathcal{H} \oplus \mathfrak{K}$, i.e. if $\{x_n\} \subseteq D(S)$, $x_n \to x$, and $Sx_n \to y$ imply $x \in D(S)$ and $Sx = y$. If there is a closed operator T such that $T \supset S$ (i.e. $G(S) \subset G(T)$), then S is said to be closable. This is the case if and only if the closure of G(S) is the graph of an operator, i.e. contains nothing of the form $0 \oplus y$, $y \neq 0$. In this case the operator with this graph is called the closure of S;

it is the smallest closed extension. We denote the collection of closed, densely defined operators from $\mathcal{H}$ to $\mathcal{K}$ by $\mathcal{C}(\mathcal{H}, \mathcal{K})$, and set $\mathcal{C}(\mathcal{H}) = \mathcal{C}(\mathcal{H}, \mathcal{H})$.

When S is densely defined, the <u>adjoint</u> $S^*: \mathcal{K} \to \mathcal{H}$ is defined by letting $D(S^*)$ be the set of $y \in K$ such that the linear functional $x \to (Sx, y)$ is bounded with respect to the $\mathcal{H}$-norm on $D(S)$. If so, there is a unique element $S^*y \in H$ such that

$$(Sx, y) = (x, S^*y) \quad , \quad \text{all } x \in D(S).$$

Let $J: \mathcal{H} \oplus \mathcal{K} \to \mathcal{K} \oplus \mathcal{H}$ be defined by $J(x \oplus y) = -y \oplus x$. Then it is easily seen that the graph of S^* is given by

$$G(S^*) = [JG(S)]^{\perp} . \tag{2.1}$$

PROPOSITION 2.1. Suppose $S: \mathcal{H} \to \mathcal{K}$ is densely defined. Then S^* is closed. Moreover, S^* is densely defined if and only if S is closable, and if so the closure is $(S^*)^*$.

<u>Proof</u>. Equation (2.1) shows that $G(S^*)$ is closed, so S^* is closed. There is no element $0 \oplus y$, $y \neq 0$ in the closure of $G(S)$ if and only if $\{ z \mid z \oplus x \perp J(G(S))\}$ is dense in $\mathcal{K}$, i.e. $D(S^*)$ is dense. If so, the closure of $G(S)$ is $[J(G(S))]^{\perp\perp} = [JG(S^*)]$ $= G(S^{**})$. ∎

The adjoint operator has the properties

$$(\alpha S)^* = \bar{\alpha} S^* , \quad \alpha \neq 0 ; \tag{2.2}$$
$$S^* + T^* \subset (S+T)^* \quad ; \quad S^*T^* \subset (TS)^* .$$

In general the inclusions are strict. An important exception is the case $T \in \mathcal{B}(\mathcal{H}, \mathcal{K})$. Then $D(S+T) = D(S)$ and likewise

$D(S^*) = D(S^* + T^*) \subset D((S+T)^*)$, while

$D((S+T)^*) = D((S+T)^* - T^*) \subset D(S^* + T^* - T^*) = D(S^*)$. Thus

$D((S+T)^*) = D(S^*) = D(S^* + T^*)$, so $(S+T)^* = S^* + T^*$.

An operator $T \in \mathcal{B}(\mathcal{K}, \mathcal{H})$ is said to be an <u>inverse</u> for $S: \mathcal{H} \to \mathcal{K}$ if

$$ST = I, \quad TS \subset I, \tag{2.3}$$

the identity operators in $\mathcal{K}$ and $\mathcal{H}$ respectively. These imply that S and T are 1-1, $R(S) = \mathcal{K}$, $R(T) = D(S)$. Moreover, $G(T) = \{y \oplus x \mid x \oplus y \in G(S)\}$, so T is uniquely determined. Also, T is closed, so S is necessarily closed. Conversely, suppose S is closed, 1-1, and onto. The unique operator T satisfying (2.3) then is closed and everywhere defined. By the closed graph theorem it is then bounded.

The <u>resolvent set</u> $\rho(S)$ of $S: \mathcal{H} \to \mathcal{H}$ is again defined to be the set of complex λ such that $\lambda - S \equiv \lambda I - S$ has an inverse, and the <u>resolvent operator</u> is $(\lambda - S)^{-1}$. The <u>spectrum</u> is the complement $\sigma(S)$ of $\rho(S)$ in $\mathbb{C}$. If $\lambda \in \rho(S)$, then $\lambda - S$ is closed, so S is closed. Conversely the above remarks show that if S is closed then $\lambda \in \rho(S)$ if and only if $\lambda - S$ is 1-1 and onto. Even for closed operators either $\sigma(S)$ or $\rho(S)$ may be empty, but the other conclusions of Proposition 1.2 remain valid: $R(S)$ is open and $\lambda \to (\lambda - S)^{-1}$ is holomorphic. The proofs are unchanged.

Suppose $\lambda, \mu \in \rho(S)$. Then

$$(\lambda - S)^{-1} - (\mu - S)^{-1} = (\mu - \lambda)(\lambda - S)^{-1}(\mu - S)^{-1}. \tag{2.4}$$

This is called the (first) <u>resolvent equation</u> and is proved by multiplying the left side by $(\lambda - S)^{-1}(\lambda - S)$ on the left and by $(\mu - S)(\mu - S)^{-1}$ on the right. It follows that $(\lambda - S)^{-1}$ and $(\mu - S)^{-1}$ commute.

If $S \in \mathcal{C}(\mathcal{H})$ and $\lambda \in \rho(S)$, then $\bar{\lambda} \in \rho(S^*)$ and

$$(\bar{\lambda} - S^*)^{-1} = [(\lambda - S)^{-1}]^*. \tag{2.5}$$

To see this, note first that $(\lambda - S)^* = \bar{\lambda} - S^*$, since λI is bounded. We may replace S by $S - \lambda$ and take $\lambda = 0$. Let $T = S^{-1}$. Direct computation shows $R(T^*) \subset D(S^*)$ and $S^*T^* = I$, $T^*S^* \subset I$. Thus $T^* = (S^*)^{-1}$.

2. Symmetric and self-adjoint operators

An operator $S: \mathcal{H} \to \mathcal{H}$ is said to be symmetric if $(Sx, y) = (x, Sy)$, all $x, y \in D(S)$. This is equivalent, by polarization, to (Sx, x) real, all $x \in D(S)$. If S is densely defined, it is symmetric if and only if $S \subset S^*$. If $S = S^*$, S is said to be self-adjoint. The following criterion is useful.

PROPOSITION 2.2. Let $S: \mathcal{H} \to \mathcal{H}$ be densely defined and symmetric. The following are equivalent:

(a) $S = S^*$

(b) $\sigma(S) \subset \mathbb{R}$

(c) $+i, -i \in \rho(S)$.

Proof. Suppose (a) and suppose $\lambda = \mu + i\nu$ with μ, ν real and $\nu \neq 0$. Exactly as in the proof of Proposition 1.3, $\|(\lambda - S)x\| \geq \nu\|x\|$, $x \in D(S)$. Therefore $\lambda - S$ is 1-1 and has closed range (since $S = S^*$ is closed). Again $R(\lambda - S)^{\perp} = N((\lambda - S)^*)$ $= N(\overline{\lambda} - S) = (0)$, so $\lambda - S$ is onto. Thus $\lambda \in \rho(S)$.

Clearly (b) implies (c). Suppose $\pm i \in \rho(S)$. Then $N(S^* + i) = R(S^* - i)^{\perp} = (0)$. This means that $S^* + i$, which is an extension of the onto operator $S + i$, is 1-1. This is only possible if $S^* + i = S + i$, which implies $S^* = S$. ∎

For the remainder of this section, let S be a self-adjoint operator in $\mathcal{H} \neq (0)$. Note that $\sigma(S)$ is not empty. In fact, let $T = (S + i)^{-1}$. By (2.4) and (2.5), T is normal. Since T is 1-1, it is not 0. By (1.7), there is a $\lambda \neq 0$, $\lambda \in \sigma(T)$. It is easily seen that $\lambda - T = \lambda(S + i - \lambda^{-1})T$. Since $\lambda - T$ is either not 1-1 or not onto, and since $R(T) = D(S)$, this implies that $\lambda^{-1} - i \in \sigma(T)$.

Let $\tilde{\sigma}(S) = \sigma(S)$ if $\sigma(S)$ is bounded, and let $\tilde{\sigma}(S)$ be the one-point compactification of S otherwise. Let $C(\tilde{\sigma}(S))$ be the algebra of continuous complex-valued functions on $\tilde{\sigma}(S)$; thus if $\sigma(S)$ is unbounded, it consists of those continuous $f: \sigma(S) \to \mathbb{C}$ such that $\lim_{|s| \to \infty} f(s)$ exists. This algebra has a norm $\|f\| = \sup_{s \in \sigma(S)} |f(s)|$ and an adjoint operation $f^*(s) = \overline{f(s)}$. It contains the functions

(2.6) $$e(s) \equiv 1, \quad r_{\lambda}(s) = (\lambda - s)^{-1}, \quad \lambda \in \rho(S).$$

Corresponding to these functions are the operators I and $R_{\lambda} = (\lambda - S)^{-1}$. We want to extend this correspondence.

THEOREM 2.1. Let $S: \mathcal{H} \to \mathcal{H}$ be self-adjoint, $\mathcal{H} \neq (0)$. The mapping $e \to I$, $r_{\lambda} \to (\lambda - S)^{-1}$ has a unique extension to an isometric algebra isomorphism $f \to f(S)$ from $C(\tilde{\sigma}(S))$ to $\mathcal{B}(\mathcal{H})$.

This mapping preserves adjoints and order: $f(S)^* = f^*(S)$, $f(S) \geq 0$ if and only if $f \geq 0$ on $\sigma(S)$.

Proof. Let $\mathcal{F}$ be the algebra of functions on $\mathbb{C}$ generated by e and r_λ, $\lambda \in \rho(S)$. If such a mapping exists and p is a polynomial in n-variables, the function f defined by $f(s) = p(r_{\lambda_1}(s), \ldots, r_{\lambda_n}(s))$ must correspond to the operator $f(S) = p(R_{\lambda_1}, \ldots, R_{\lambda_n})$. Conversely this will extend the mapping to $\mathcal{F}$ provided it is well-defined, i.e. provided $p(R_{\lambda_1}, \ldots, R_{\lambda_n}) = 0$ if $p(r_{\lambda_1}, \ldots, r_{\lambda_n}) \equiv 0$. To show this, we induce on n. For $n = 1$, $p(r_{\lambda_1}) \equiv 0$ implies p is the zero-polynomial. Suppose our assertion is true in $\leq m$ variables and suppose $p(r_{\lambda_1}, \ldots, r_{\lambda_{m+1}}) \equiv 0$. Since each $r_{\lambda_j} = 0$ at ∞, the constant term of p is zero. Now

$$(\lambda_{m+1} - s)r_{\lambda_j}(s) \equiv (\lambda_{m+1} - \lambda_j)r_{\lambda_j}(s) + 1,$$

so $0 \equiv (\lambda_{m+1} - s)p(r_{\lambda_1}, \ldots, r_{\lambda_{m+1}}) = q(r_{\lambda_1}, \ldots, r_{\lambda_m})$. But also

$$(\lambda_{m+1} - S)R_{\lambda_j} = (\lambda_{m+1} - \lambda_j)R_{\lambda_j} + I,$$

so $(\lambda_{m+1} - S)p(R_{\lambda_1}, \ldots, R_{\lambda_{m+1}}) = q(R_{\lambda_1}, \ldots, R_{\lambda_m}) = 0$ by the induction assumption. Now p has no constant term, each R_{λ_j} has range $D(S)$, and $\lambda_{m+1} - S$ is 1-1, so this implies that $p(R_{\lambda_1}, \ldots, R_{\lambda_{m+1}}) = 0$. Thus there is a unique extension to an algebra homomorphism $f \to f(S)$ from $\mathcal{F}$ to $\mathcal{B}(\mathcal{H})$.

Next, for $f \in \mathcal{F}$, $\sigma(f(S)) = f(\tilde{\sigma}(S))$. To see this, note that $\mathcal{F}$ consists precisely of all bounded rational functions which are holomorphic on $\sigma(S)$. Suppose $\mu \notin f(\tilde{\sigma}(S))$. Then $(\mu - f(s))^{-1} = h(s) \in \mathfrak{F}$ and

$$(\mu - f(S))h(S) = e(S) = I = h(S)(\mu - f(S)).$$

Thus $\mu \notin \sigma(f(S))$. Conversely if $\lambda \in \sigma(S)$ and $\mu = f(\lambda)$, then $\mu - f(s) = (\lambda - s)g(s)$, where $g \in \mathfrak{F}$ and $g(\infty) = 0$. Therefore, using the partial fractions decomposition of g we get

$$g(S)(\lambda - S) \subset (\lambda - S)g(S) = \mu - f(S).$$

Since $\lambda - S$ is either not 1-1 or not onto, the same is true of $\mu - f(S)$. Thus $\sigma(f(S))$ contains $f(\sigma(S))$ and hence its closure $f(\tilde{\sigma}(S))$.

From (1.7) and what has just been proved we get

$$\|f(S)\| = \sup\{f(\lambda) \mid \lambda \in \tilde{\sigma}(S)\} = \|f\|, \quad f \in \mathfrak{F}.$$

Therefore this map is an isometry of $\mathfrak{F}$, considered as a subalgebra of $C(\tilde{\sigma}(S))$, into $\mathcal{B}(H)$. On $\sigma(S) \subseteq \mathbb{R}$, $r_\lambda^* = r_{\bar{\lambda}}$. Therefore $\mathfrak{F}$ is a closed subalgebra which separates points and is closed under complex conjugation. The Stone-Weierstrass theorem shows that $\mathfrak{F}$ is dense in $C(\tilde{\sigma}(S))$, so $f \rightarrow f(S)$ has a unique extension to an algebra isometry. By (2.5), $r_\lambda^*(S) = r_{\bar{\lambda}}(S) = r_\lambda(S)^*$. Thus $f^*(S) = f(S)^*$, $f \in \mathfrak{F}$, so this carries over to $f \in C(\tilde{\sigma}(S))$.

Finally, suppose $f \geq 0$, $f \in C(\tilde{\sigma}(S))$. Then $f = g^2$, where $g \in C(\tilde{\sigma}(S))$, $g \geq 0$. Then $f(S) = g(S)^2 = g(S)^* g(S) \geq 0$. Conversely if $f(S) \geq 0$, then $f(\tilde{\sigma}(S)) = \sigma(f(S)) \subseteq [0, \infty)$, by Proposition 1.3. Thus $f \geq 0$. ∎

COROLLARY. Suppose S is self-adjoint. Then S is bounded if and only if $\sigma(S)$ is bounded.

Proof. We know $\sigma(S)$ bounded if S bounded. Conversely, suppose $\sigma(S)$ bounded. Let $f(s) \equiv s$. Then $f \in C(\sigma(S))$. Moreover, $(f - \lambda)r_\lambda = e$, $\lambda \in \rho(S)$, so $(f(S) - \lambda)R_\lambda = R_\lambda(f(S) - \lambda) = I$. Therefore $f(S) - \lambda = S - \lambda$, so $S = f(S) \in \mathcal{B}(\mathcal{H})$. ∎

To illustrate Theorem 2.1, suppose $\mathcal{H}$ is finite-dimensional. Then S is automatically bounded, and $\sigma(S)$ consists of the eigenvalues $\lambda_1 < \lambda_2 < \ldots < \lambda_n$ of S. Every function on $\sigma(S)$ is continuous. Let p_j be the function with $p_j(\lambda_k) = \delta_{jk}$, and set $P_j = p_j(S)$. Then $p_j = p_j^2 = p_j^*$, so P_j is a projection. Also, $P_jP_k = \delta_{jk}I$ and $P_1 + \ldots + P_n = I$. Thus the $\mathcal{H}_j = P_j(\mathcal{H})$ are orthogonal subspaces with $\bigvee \mathcal{H}_j = \mathcal{H}$. For any function f on $\sigma(S)$, $f = \sum f(\lambda_j)p_j$. Therefore $f(S) = \sum f(\lambda_j)P_j$, and $f(S)P_j = P_jf(S) = f(\lambda_j)P_j$. In particular, taking $f(s) \equiv s$, we get $SP_j = P_jS = \lambda_jP_j$. Thus $\mathcal{H}_j$ is invariant for S and $Sx = \lambda_jx$, $x \in \mathcal{H}_j$. We have shown, in effect, that $\mathcal{H}$ is spanned by the eigenspaces of S and that these spaces are orthogonal. Equivalently, there is an orthonormal basis for $\mathcal{H}$ such that the matrix of S with respect to this basis is diagonal; the matrix of f(S) is the diagonal matrix obtained by applying f to the entries of the matrix for S.

3. Extending the functional calculus

The order preserving property of the isomorphism $f \to f(S)$ is the basis for further extensions of this mapping. In this section we shall ignore the genesis of this isomorphism and extend it in two steps. The first is to an algebra of bounded but not necessarily continuous functions, the second is to an algebra of functions not necessarily bounded or continuous.

Let σ be a closed subset of $\mathbb{R}$, and let $\mathfrak{A}_0 = \mathfrak{A}_0(\sigma)$ be the algebra of continuous complex-valued functions on σ which have limit 0 at ∞ ; if σ is bounded this is just $C(\sigma)$. Let $\mathfrak{A}_+ = \mathfrak{A}_+(\sigma)$ be the set of bounded functions on σ which are the pointwise limits of increasing sequences of non-negative functions in $\mathfrak{A}_0$. Let $\mathfrak{A}_1$ be the algebra of functions on σ generated by $\mathfrak{A}_+$. Then $\mathfrak{A}_+$ contains all bounded continuous non-negative functions on σ, so $\mathfrak{A}_1$ contains all bounded continuous functions on σ, whether or not they have limits at ∞. Write $f_n \nearrow f$ if f_n is real-valued and $f_n(s)$ increases to $f(s)$ as $n \to \infty$, all $s \in \sigma$.

Let $\mathcal{H} \neq (0)$ be a Hilbert space. A collection $\{S\} \subseteq \mathcal{B}(\mathcal{H})$ will be called <u>full</u> if $\bigvee R(S) = \mathcal{H}$. Note that if $S = S^*$, then $R((S+i)^{-1}) = D(S)$ is dense in $\mathcal{H}$, so $\{(\lambda - S)^{-1}\}$ is full.

PROPOSITION 2.3. Let σ be a closed subset of $\mathbb{R}$, and let $A_0 : \mathfrak{A}_0(\sigma) \to \mathcal{B}(\mathcal{H})$ be an isometric algebra isomorphism which preserves adjoints. Then A_0 has a unique extension to an algebra homomorphism $A_1 : \mathfrak{A}_1(\sigma) \to \mathcal{B}(\mathcal{H})$ with the property that if $\{f_n\} \subset \mathfrak{A}_+$ and $f_n \nearrow f \in \mathfrak{A}_+$, then $A_1(f_n) \to A_1(f)$. The extension A_1 also preserves adjoints and order, and $\|A_1(f)\| \leq \sup_\sigma |f(s)|$. The image of $\mathfrak{A}_o$ is full if and only if $A_1(e) = I$, where $e(s) \equiv 1$.

Proof. The argument in Theorem 2.1 shows that A_o preserves order. Suppose $0 \leq f_n \uparrow f$, where $f_n \in \mathfrak{A}_o$, f bounded. Then $\{A_o(f_n)\}$ is a bounded increasing sequence of operators in $\mathcal{B}(\mathcal{H})$, so $A_o(f_n) \to S \in \mathcal{B}(\mathcal{H})$. Suppose also $g_n \in \mathfrak{A}_o$, $0 \leq g_n \uparrow f$. Then $A_o(g_n) \to T \in \mathcal{B}(\mathcal{H})$. Let $f_n \wedge g_m(s) = \min\{f_n(s), g_m(s)\}$. Then $f_n \wedge g_m \uparrow f_n$ as $m \to \infty$. These functions are all continuous and vanish at ∞ (if σ is unbounded), so the convergence is uniform. Therefore $T = \lim A_o(g_m) \geq \lim A_o(f_n \wedge g_m) = A_o(f_n)$, all n, so $T \geq S$. Also $S \leq T$, so the limit $S = T = A_1(f)$ is independent of the approximating sequence. If $f \in \mathfrak{A}_o \cap \mathfrak{A}_+$, set $f_n \equiv f$. Then $A_1(f) = \lim A_o(f_n) = A_o(f)$.

Suppose $f_n, g_n \in \mathfrak{A}_o$, $0 \leq f_n \uparrow f \in \mathfrak{A}_+$, $0 \leq g_n \uparrow g \in \mathfrak{A}_+$. Then $f_n + g_n \uparrow f+g$, $f_n g_n \uparrow fg$, so $A_1(f+g) = A_1(f) + A_1(g)$, $A_1(fg) = A_1(f)A_1(g)$. Similarly, if $f \in \mathfrak{A}_+$, $\alpha \geq 0$, then $A_1(\alpha f) = \alpha A_1(f)$. It is easily seen that if $f \leq g$, $f, g \in \mathfrak{A}_+$, then $A_1(f) \leq A_1(g)$. Now since $\mathfrak{A}_+$ is closed under addition, multiplication, and multiplication by non-negative reals, $\mathfrak{A}_1$ consists precisely of all functions of the form $f = f_1 - f_2 + if_3 - if_4$, $f_j \in \mathfrak{A}_+$. If also $f = g_1 - f_2 + ig_3 - ig_4$, $g_j \in \mathfrak{A}_+$, then $f_1 + g_2 = g_1 + f_2$, $f_3 + g_4 = g_3 + f_4$. It follows from the properties of A_1 that

$$A_1(f_1) - A_1(f_2) + iA_1(f_3) - iA_1(f_4) = A_1(g_1) - A_1(g_2) + iA_1(g_3) - iA_1(g_4).$$

Therefore A_1 has a unique linear extension to $\mathfrak{A}_1$, and this extension is seen to be an algebra homomorphism. Moreover, $A_1(f^*) = A_1(f)^*$. If $f, g \in \mathfrak{A}$ and $f \geq g$, then $f = f_1 - f_2$, $g = g_1 - g_2$, $f_j, g_j \in \mathfrak{A}_+$ and $f_1 + g_2 \geq g_1 + f_2$, so $A_1(f_1 + g_2) \geq A_1(g_1 + f_2)$, so $A(f) \geq A(g)$.

If $f_n \in \mathfrak{A}_+$ and $f_n \uparrow f \in \mathfrak{A}_+$, take sequences $\{f_{n,m}\} \subset \mathfrak{A}_o \cap \mathfrak{A}_+$ with $f_{n,m} \uparrow f_n$. Let $g_n(s) = \max\{f_{1,n}(s), \ldots, f_{n,n}(s)\}$. Then $g_n \in \mathfrak{A}_o \cap \mathfrak{A}_+$ and $g_n \leq f_n$, $g_n \uparrow f$. Therefore $A_1(g_n) \leq A_1(f_n) \leq A_1(f)$, and since $A_1(g_n) \to A_1(f)$ we have $A_1(f_n) \to A_1(f)$. If $|f(s)| \leq c$, then $(f^*f)(s) \leq c^2$, so $A_1(f)^*A_1(f) \leq c^2 A_1(e) \leq c^2 I$, so $\|A_1(f)\|^2 \leq c^2$. Finally, if $\{A_o(f)\}$ is full, then since $A_1(e)A_1(f) = A_1(ef) = A_1(f)$, all f, we have $A_1(e) = I$. Conversely, let $\mathcal{H}_1 = \bigvee R(A_o(f))$. Clearly each $A_1(f)$ has range in $\mathcal{H}_1$. Therefore $A_1(e) = I$ implies fulness.

Let e^N be the characteristic function of the interval $(-N,N)$. The restriction of e^N to σ, also denoted by e^N, is in $\mathfrak{A}_+$. Let $\mathfrak{A}$ be the algebra of functions on K which are locally in $\mathfrak{A}_1$ i.e., those f such that $e^N f \in \mathfrak{A}_1$, all $N > 0$. Put $f^N = e^N f$.

THEOREM 2.2. Let σ be a closed subset of $\mathbb{R}$ and let $A_o: \mathfrak{A}_o(\sigma) \to \mathcal{B}(\mathcal{H})$ be an isometric algebra isomorphism onto a full subalgebra of $\mathcal{B}(\mathcal{H})$ of which preserves adjoints. There is unique extension $A: \mathfrak{A}(\sigma) \to \mathcal{C}(\mathcal{H})$, the space of densely defined, closed operators in $\mathcal{H}$, which has the properties:

(i) $A(\alpha f) = \alpha A(f)$, $\alpha \neq 0$;

(ii) $A(f) + A(g) \subset A(f+g)$;

(iii) $A(f)A(g) \subset A(fg)$;

(iv) $A(f^*) = A(f)^*$;

(v) f bounded $\Longrightarrow \|A(f)\| \leq \sup_\sigma |f(s)|$;

(vi) $f \geq 0 \Longrightarrow (A(f)x, x) \geq 0$, all $x \in D(A(f))$;

(vii) $f_n \in \mathfrak{A}_+$, $f_n \nearrow f \Longrightarrow (A(f_n)x, x) \to (A(f)x, x)$; all $x \in D(A(f))$;

(viii) if $x \in D(A(f))$, then $A(f^N)x \to A(f)x$ as $N \to \infty$.

Proof. The preceding Proposition gives us a unique extension $A_1: \mathfrak{A}_1 \to \mathcal{B}(\mathcal{H})$. Given $f \in \mathfrak{A}$, let $D(A(f)) = \{x \mid \sup_N \|A_1(f^N)x\| < \infty\}$. Suppose $x \in D(A(f))$. We have $\|A_1(f^N)x\|^2 = (A_1(|f^N|^2)x, x)$ bounded and non-decreasing with N, hence convergent as $N \to \infty$. For $M < N$, $(f^*)^M f^N = (f^*)^M f^M$, so

$$\|A_1(f^N)x - A_1(f^M)x\|^2 = \|A_1(f^N)x\|^2 - \|A_1(f^M)\|^2 \to 0$$

as $M, N \to \infty$. Let $A(f)x$ be the limit. The operator $A(f)$ so defined is linear. Since $e^N \nearrow e$, $A_1(e^N) \to A_1(e) = I$. Therefore if $f \in \mathfrak{A}_1$, so that $D(A(f)) = \mathcal{H}$, we have $A(f) = \lim A_1(f^N) = \lim A_1(e^N)A_1(f) = A_1(f)$. In general, if f is bounded then the f^N are uniformly bounded; by the Proposition $\|A_1(f^N)\| \leq \sup |f^N(s)| \leq \sup |f(s)|$. Therefore $A(f) \in \mathcal{B}(\mathcal{H})$, $\|A(f)\| \leq \sup |f(s)|$.

Each $E^N = A(e^N)$ is a projection whose range is clearly contained in $D(A(f))$ for any $f \in \mathfrak{A}$. Since $E^N \to I$, $\bigvee E^N(\mathcal{H}) = \mathcal{H}$. Therefore each $A(f)$ is densely defined. Suppose $y \in D(A(f^*))$, $x \in D(A(f))$. Then

$$(y, A(f)x) = \lim(y, A(f^N)x) = \lim(A(f^*)^N)x, y) = (A(f^*)y, x).$$

Conversely, suppose $y \in D(A(f)^*)$ and $A(f)^*y = z$. Then

$$(E^N y, A(f^N)x) = (y, A(f)E^N x) = (z, E^N x) = (E^N z, x).$$

Thus $A((f^*)^N)y = E^N z \to z$, so $y \in D(A(f^*))$ and $A(f^*)y = z$. This proves (iv) and it follows that $A(f) = A(f^*)^*$ is closed.

So far we have defined A, shown that $A(f) \in \mathcal{C}(\mathcal{H})$ and proved (iv) and (viii). Before turning to the other properties, let us show uniqueness. If B is a second extension of A_o, it also extends A_1. By (viii), $B(f) \subset A(f)$. Taking adjoints, $B(f) = B(f^*)^* \supset A(f^*)^* = A(f)$. Therefore $B(f) = A(f)$.

Property (i) is trivial. Suppose $f, g \in \mathfrak{A}$. Then $(f+g)^N = f^N + g^N$, $(fg)^N = f^N g^N$. Then $\| A_1((f+g)^N)x \| \leq \| A_1(f^N)x + A_1(g^N)x \|$, so $D(A(f)) \cap D(A(g)) \subset D(A(f+g))$. For $x \in D(A(f)) \cap D(A(g))$,

$$A(f+g)x = \lim (A_1((f+g)^N x) = \ldots = A(f)x + A(g)x.$$

If $x \in D(A(g))$ and $A(g)x \in D(A(f))$, then $\| A((fg)^N)x \| = \| A(f^N)A(g)x \|$ is bounded. Thus $D(A(f)A(g)) \subset D(A(fg))$ and $A(f)A(g)x = \lim A_1(f^N)A(g)x = \lim A_1(f^N)(\lim A_1(g^M)x) = \lim A_1((fg)^N x)$ $= A(fg)x$, since $A_1(f^N)A_1(g^M) = A_1((fg)^N)$, $N < M$.

If $f \geq 0$ then $f^N \geq 0$,, so (vi) follows from the fact that A_1 preserves order. Finally, suppose $f_n \in \mathfrak{A}_+$, $f_n \uparrow f \in \mathfrak{A}$. Then for $x \in D(A(f))$,

$$(A(f_n^N)x, x) \leq (A(f_n)x, x) \leq (A(f)x, x) .$$

By the preceding Proposition, $A(f_n^N) \to A(f^N)$, and since $(A(f^N)x, x) \to (A(f)x, x)$, we get (vii). ∎

COROLLARY. Suppose S is self-adjoint and $\sigma = \sigma(S)$. Let $A_0 : \mathfrak{A}_0(\sigma) \to \mathcal{B}(\mathcal{H})$ be given by $A_0(f) = f(S)$. Let A be the extension of the preceding theorem, and let $f(s) \equiv s$. Then $A(f) = S$.

Proof. Let $h(s) = (s - i)^{-1}$. Then $hf = 1 + if$, so

$$(S - i)^{-1} A(f) \subset I + i(S - i)^{-1} .$$

Multiplication on the left by $S - i$ gives $A(f) \subset S$. Then $A(f) = A(f)^* \supset S^* = S$, so $A(f) = S$. ∎

Combining Theorems 2.1 and 2.2 as in the corollary, we have a mapping from $\mathfrak{A}(\sigma(S))$ to $\mathcal{C}(\mathcal{H})$ which we shall still denote by $g \longrightarrow g(S)$. (For a more explicit construction of this mapping, both in general and for an important special case, see §3 of the next chapter.)

3

SPECTRAL MEASURES AND UNITARY GROUPS

1. Correspondence between self-adjoint operators, spectral measures, and unitary groups

Let S be a self-adjoint operator in $\mathcal{H} \neq (0)$. Let e_t be the function on $\mathbb{R}$, or its restriction to $\sigma(S)$, defined by $e_t(s) = 1$, $s \leq t$, $e_t(s) = 0$, $s > t$. These functions are bounded and are in $\mathfrak{A}(\sigma(S))$. Again let $e(s) \equiv 1$. Let $f_t(s) = 1$, $s < t$, $f_t(s) = 0$, $s \geq t$. Then it is easy to see that $e_t = e_t^* = e_t^2$, $e_s \leq e_t$ if $s \leq t$, $e - e_t \in \mathfrak{A}_+$ and $e - e_t \nearrow e - e_s$ if $t \searrow s$, $f_t \in \mathfrak{A}_+$ and $f_t \leq e_t \leq e$ and $f_t \nearrow e$ as $t \to +\infty$, while $e - e_t \nearrow e$ as $t \to -\infty$. If $E_t = e_t(S)$, the family of operators $\{E_t\}_{t \in \mathbb{R}}$ has the corresponding properties:

each E_t is a projection;

$E_s \leq E_t$ if $s \leq t$;

$E_t \to E_s$ as $t \searrow s$;

$E_t \to I$ as $t \to +\infty$, $E_t \to 0$ as $t \to -\infty$.

Such a family $\{E_t\}$ is called a <u>spectral measure</u> (or a resolution of the identity).

Conversely, suppose $\{E_t\}$ is a spectral measure in $\mathcal{H}$. The <u>support</u> is the complement in $\mathbb{R}$ of the union of all open intervals (a, b) such that $E_b = E_a$. We shall show that $\{E_t\}$ determines a self-adjoint operator S such that $E_t = e_t(S)$. The procedure is to extend the mapping $e_t \to E_t$ and to let S correspond to the function $f(s) = s$.

PROPOSITION 3.1. Let $\{E_t\}$ be a spectral measure in with support σ. The mapping $e_t \to E_t$ has a unique extension, written $f \to \int f(s)\, dE_s$, mapping $\mathfrak{A}(\sigma)$ into $\mathcal{L}(\mathcal{H})$ and satisfying the conditions of Theorem 2.2.

Proof. If B is a half-open interval $(a, b]$, let e^B be the characteristic function. Then $e^B = e_b - e_a$. Let $E(B) = E_b - E_a$, and let $E(\emptyset) = 0$. If B_1, B_2 are two such intervals, then the fact that E_t increases with t implies $E(B_1 \cap B_2) = E(B_1)E(B_2)$, while $E(B_1 \cup B_2) = E(B_1) + E(B_2) - E(B_1 \cap B_2)$, if $B_1 \cup B_2$ is again an interval. It follows that the map $e^B \to E(B)$ has a unique linear extension which is an algebra homomorphism from the algebra of functions generated by the e^B into $\mathcal{B}(\mathcal{H})$. Furthermore if $f = \Sigma\, \alpha_j e^{B_j}$ where the B_j are disjoint (and the sum is finite), then for $x \in \mathcal{H}$,

$$\begin{aligned} \|\Sigma\, \alpha_j E(B_j)x\|^2 &= \Sigma\, |\alpha_j|^2 \|E(B_j)x\|^2 \\ &\leq \sup\{|\alpha_j| \,|\, E(B_j) \neq 0\} \cdot \Sigma\, \|E(B_j)x\|^2 \\ &\leq \sup_\sigma\{|f(s)|\} \cdot \|x\|^2 . \end{aligned}$$

Conversely, $\sup_\sigma \{|f(s)|\} = |\alpha_k|$ for some k such that $E(B_k) \neq 0$. Take x in the range of $E(B_k)$ with $\|x\| = 1$. Then $\|\Sigma\, \alpha_j E(B_j)x\|^2 = \|\alpha_k x\|^2 = |\alpha_k|^2$. Thus the map $f \to \Sigma\, \alpha_j E(B_j)$ has a unique extension to an algebra isometry into $\mathcal{B}(\mathcal{H})$ from the algebra $\mathfrak{A}_*(\sigma)$

which is the closure in the sup norm on σ of the algebra generated by the e_B. Moreover this mapping is seen to preserve adjoints. Since $\mathfrak{A}_o(\sigma) \subset \mathfrak{A}_*(\sigma)$ we may restrict to $\mathfrak{A}_o(\sigma)$ and apply Proposition 2. 3 to get a mapping $A_1: \mathfrak{A}_1(\sigma) \to \mathcal{B}(\mathcal{H})$. We assert that $A_1(e_t) = E_t$. Given $t \in \mathbb{R}$, choose a sequence $\{f_n\}$ of continuous functions with $0 \le f_n \le e$, $f_n \nearrow$, $f_n(s) = 0$, $s \le t$, $f_n(s) = 1$, $t + \frac{1}{n} \le s \le t+n$, and $f_n(s) = 0$, $s \ge t+n+1$. Then $f_n \nearrow 1 - e_t$. Let $B_n = (t + \frac{1}{n}, f+n]$, and let $F_n = A_1(f_n)$. Our construction gives $E(B_n) \le F_n \le I - E_t$, clearly. But $E(B_n) \to I - E_t$, so $A_1(1 - e_t)$ $= \lim A_1(f_n) = I - E_t$, or $A_1(e_t) = E_t$. Thus A_1 extends the map $e_t \to E_t$; moreover $E_t \nearrow I$ as $t \to \infty$ implies that $A_1(\mathfrak{A}_o)$ is full. The assertion now follows from Theorem 22. ∎

Now given a spectral measure $\{E_t\}$, let

$$U_t = \int e^{ist} \, dE_s .$$

Then U_t corresponds to the function $u_t(s) = e^{ist}$. The family of operators $\{U_t\}_{t \in \mathbb{R}}$ has the following properties:

each U_t is unitary;

$U_{s+t} = U_s U_t$, $s, t \in \mathbb{R}$;

$U_t \to U_s$ as $t \to s$.

The first two properties follow immediately from the corresponding properties of the u_t: $u_t^* u_t = 1$, $u_{s+t} = u_s u_t$. To prove the third property, suppose x is in the range of $E_b - E_a$. Now $u_t(e_b - e_a) \to u_s(e_b - e_a)$ uniformly as $t \to s$, so $U_t x \to U_s x$ as $t \to s$. Since $\bigvee R(E_b - E_a) = \mathcal{H}$, the assertion follows from the following lemma, whose proof is left to the reader.

LEMMA. Let $\{S_n\} \subset \mathcal{B}(\mathcal{H})$, $S \in \mathcal{B}(\mathcal{H})$. Suppose $S_n x \to Sx$ for each x in a dense subset of $\mathcal{H}$, and suppose $\|S_n\| \leq M$ all n, for some constant M. Then $S_n \to S$.

A collection of operators $\{U_t\}_{t \in \mathbb{R}}$ satisfying the above conditions is called a <u>strongly-continuous one-parameter unitary group</u>; to be brief, if somewhat incorrect, we shall simply call it a <u>unitary group</u>.

If S is self-adjoint and $E_t = e_t(S)$, then the uniqueness of the extension of $e_t \to E_t$ implies that for any $f \in \mathfrak{A}(\sigma(S))$, $f(S) = \int f(s)dE_s$. In particular, $U_t = \int e^{ist} dE_s = u_t(S)$. Conversely, let $\{U_t\}$ be a unitary group. We want to find a self-adjoint S such that $U_t = u_t(S)$. Heuristically, $U_t = \exp(itS)$ and $S = \frac{1}{i}\frac{d}{dt} U_t \Big|_{t=0}$. This turns out actually to work.

Given a unitary group $\{U_t\}$ define

$$S = \lim_{t \to 0} \frac{1}{it}(U_t - I)$$

to be the operator with domain $D(S)$ the set of all x such that $y = \lim_{t \to 0} \frac{1}{it}(U_t x - x)$ exists, and let $Sx = y$.

PROPOSITION 3.2. Let $\{U_t\}$ be a unitary group in $\mathcal{H}$ and let $S = \lim_{t \to 0} \frac{1}{it}(U_t - I)$. Then S is self-adjoint.

<u>Proof.</u> Suppose φ is a continuous function on $[0, \infty)$ with $\int_0^\infty |\varphi(s)| ds < \infty$. Define $T_\varphi x$, $x \in \mathcal{H}$, by

$$T_\varphi x = \int_0^\infty \varphi(s) U_s x \, ds.$$

Since $U_t x$ is continuous and has norm $\equiv \|x\|$, this exists as an improper Riemann integral. Clearly $T_\varphi \in \mathcal{B}(\mathcal{H})$. Suppose also that the derivative φ' is continuous and absolutely integrable on $[0, \infty)$. For $t > 0$,

$$\frac{1}{it}(U_t - I)T_\varphi x = \frac{1}{it}\int_0^\infty \varphi(s)[U_{t+s}x - U_s x]ds$$

$$= \int_t^\infty \frac{1}{it}[\varphi(s-t) - \varphi(s)]U_s x\, ds + \frac{1}{it}\int_0^t \varphi(s)U_s x\, ds.$$

This and a similar expression for $s < 0$ show that $T_\varphi x \in D(S)$ and

(3.1) $$ST_\varphi = iT_{\varphi'} + i\varphi(0)I ,$$

since $U_s x \to x$ as $s \to 0$. If $\{\varphi_n\}$ is a sequence of functions as above with $\int_0^\infty \varphi_n(s)ds = 1$, $\varphi_n(s) = 0$ for $s \geq \frac{1}{n}$, then it is easy to see that $T_{\varphi_n} x \to x$. It follows that $D(S)$ is dense. A routine calculation shows that S is symmetric.

We want to show that $\pm i \in \rho(S)$. Taking $\varphi(s) = ie^{-s}$ in (3.1) we get $(S + i)R_+ = I$, where

(3.2) $$R_+ x = -i\int_0^\infty e^{-s}U_s x\, ds ,$$

Since $U_t U_s = U_s U_t$, $D(S)$ is invariant under U_t and $U_t S \subset SU_t$. It follows that $R_+(S+i) \subset (S+i)R_+ = I$. Therefore $-i \in \rho(S)$ and $R_+ = (S+i)^{-1}$. A similar argument shows that $i \in \rho(S)$ and $(S-i)^{-1} = R_-$ where

(3.3) $$R_- x = i\int_{-\infty}^0 e^{s}U_s x\, dx = i\int_0^\infty e^{-s}U_{-s} x\, ds.$$

By Proposition 2.2, S is self-adjoint. ∎

THEOREM 3.1. There are 1-1 correspondences between self-adjoint operators S, spectral measures $\{E_t\}$, and unitary groups $\{U_t\}$ given by:

$$E_t = e_t(S) , \quad S = \int s\,dE_s ;$$
$$U_t = u_t(S) , \quad S = \lim_{s \to 0} \frac{1}{is} (U_s - I) ;$$
$$U_t = \int e^{ist}\,dE_s .$$

Proof. If S is self-adjoint and $E_t = e_t(S)$, then by uniqueness of the extensions to $\mathfrak{A}(\mathbb{R})$ and the corollary of Theorem 2.2, $S = f(S) = \int f(s)dE_s$, where $f(s) = s$. Conversely, if $\{E_t\}$ is a spectral measure and $S = \int s\,dE_s$, then $(\lambda - S)^{-1} = \int r_\lambda(s)dE_s$, where $r_\lambda(s) = (\lambda - s)^{-1}$, $\lambda \notin \mathbb{R}$; this follows from the properties of the map $g \to \int g(s)dE_s$. Since the $\{r_\lambda\}$ generate a dense subalgebra of $\mathfrak{A}_o(\sigma)$, we get $g(S) = \int g(S)dE_s$ for every $g \in \mathfrak{A}_o(\sigma)$ and therefore for every $g \in \mathfrak{A}_1(\sigma)$. In particular, $E_t = \int e_t(S)dE_s = e_t(S)$.

If $U_t = u_t(S)$ and $T = \lim \frac{1}{it}(U_t - I)$, let $\{E_t\}$ be the spectral measure associated with S. Now $\frac{1}{i\varepsilon}(u_\varepsilon - e)(e_b - e_a) \to f(e_b - e_a)$ uniformly as $\varepsilon \to 0$, where $f(s) = s$. Therefore the range of $E_b - E_a$, $a < b$, is in $D(T)$ and $T = S$ on this range. For any $x \in D(S)$, $x_N = (E_N - E_{-N})x \to x$ and $Tx_N = Sx_N \to Sx$ as $N \to \infty$. Therefore $S \subset T$. By self-adjointness, $S = S^* \supset T^* = T$, so $S = T$. Conversely, let $\{U_t\}$ be a unitary group, $S = \lim \frac{1}{it}(U_t - I)$, and $V_t = u_t(S)$. By what we have just proved, $S = \lim \frac{1}{it}(V_t - I)$ also. If $x, y \in D(S)$, then $U_t x, V_t y \in D(S)$, all t. Therefore the scalar function $\varphi(t) = (U_t x, V_t y)$ is differentiable, and

$$\varphi'(t) = (iSU_t x, V_t y) + (U_t x, iSV_t y) = 0.$$

Thus $(U_t x, V_t y) \equiv (U_o x, V_o y) = (x, y)$. Since $D(S)$ is dense, this implies $V_t^* U_t \equiv I$, or $V_t \equiv U_t$.

Finally, if $U_t = \int e^{ist} dE_s$ where $\{E_s\}$ is a spectral measure, let $S = \int s\, dE_s$. Then we know $U_t = u_t(S)$, and consequently $S = \lim \frac{1}{it} (U_t - I)$. Therefore $\{U_t\}$ and $\{E_t\}$ each determine S uniquely, and so each determines the other uniquely. ∎

The correspondence between spectral measures and self-adjoint operators is one of the usual forms of the <u>spectral theorem.</u> The correspondence between self-adjoint operators and unitary groups is <u>Stone's theorem.</u> The operator S associated with $\{U_t\}$ (or the multiple iS) is often called the <u>infinitesimal generator</u> of the group. The notation $U_t = e^{itS}$ or $U_t = \exp(itS)$ is common. Note that if S is bounded, the partial sums of the Taylor series expansion of e^{its} converge uniformly on $\sigma(S)$. Therefore in this case $u_t(S) = \Sigma \frac{(itS)^n}{n!}$, the series converging uniformly.

2. <u>An example</u>

Let $\mathcal{H} = L^2(\mathbb{R})$, with Lebesgue measure understood. Let $C_c(\mathbb{R})$ be the set of continuous complex valued functions f such that $f(s) = 0$ for $|s| > M = M(f)$. Let $C_c^1(\mathbb{R})$ be the subspace consisting of those f with continuous first derivatives f'. Let S_o be the operator in $\mathcal{H}$ with $D(S_o) = C_c^1(\mathbb{R})$ and $S_o f = if'$. Then S_o is densely defined and integration by parts shows it to be symmetric. Therefore S_o^* is densely defined, so S_o has a closure S. Is S self-adjoint? If so, what are the corresponding unitary group and spectral measure?

Let us give an heuristic answer to the question of what group S generates, assuming it to be self-adjoint. Formally, and with a completely unjustified (in general) application of Taylor's formula, we have

$$U_t f(s) = \exp(itS)f(s) = \Sigma \frac{(itS)^n}{n!} f(s) = \Sigma \frac{(-t)^n}{n!} f^{(n)}(s)$$
$$= f(s-t).$$

A second heuristic way of arriving at this result is to differentiate: let $g(s,t) = U_t f(s)$. Then

$$\frac{\partial}{\partial t} g(s,t) = i \lim_{\varepsilon \to 0} \frac{1}{i\varepsilon} [U_{t+\varepsilon} f(s) - U_t f(s)]$$
$$= iSU_t f(s) = -\frac{\partial}{\partial s} g(s,t) .$$

Thus the directional derivative of g along lines $s - t = \text{constant}$ vanishes, and we have $g(s,t) = g(s-t, 0) = f(s-t)$.

If U_t is <u>defined</u> by the equation $U_t f(s) = f(s-t)$, it is easily seen that U_t is unitary and $U_{s+t} = U_s U_t$. Since $U_t f \to U_s f$ in $\mathcal{H}$ as $t \to s$ for f in the dense subspace $C_c(\mathbb{R})$, $U_t \to U_s$ as $t \to s$. Thus $\{U_t\}$ is a unitary group. Clearly $S_o \subset \lim_{t\to 0} \frac{1}{it}(U_t - I) = T$. Is T the closure of S_o?

Let us try to apply Proposition 2. 2, by inverting the operators $S_o \pm i$. If $g \in D(S_o)$ and $(S_o + i)g = f$, then $g' + g = -if$. The solution of this first order equation is given by

$$g(s) = e^{-s} g(s_o) - i \int_{s_o}^{s} e^{-s+t} f(t)dt.$$

Letting $s_o \to -\infty$ we get

(3.4) $$g(s) = R_+ f(s) = -\int_{-\infty}^{s} e^{t-s} f(t)dt = -i\int_0^{\infty} e^{-t} f(s-t)dt.$$

Similarly, if $(S_o - i)g = f$ then

$$(3.5) \qquad g(s) = R_-f(s) = -i\int_{\infty}^{s} e^{s-t}f(t)dt = i\int_0^{\infty} e^{-t}f(s+t)dt .$$

(Compare (3.4) and (3.5) with (3.2) and (3.3).) If $f, g \in C_c(\mathbb{R})$ then

$$|(R_+f, g)| = |\int_0^{\infty} e^{-s} \int_{-\infty}^{\infty} f(t-s)g^*(s)dtds| \leq \|f\| \|g\|.$$

Therefore R_+ has an extension to an operator in $\mathcal{B}(\mathcal{H})$. Since e^{-t} is square-integrable on $(0,\infty)$ the formula (4.4) also defines the extension, which we still denote by R_+. Note that (3.4) shows that the range of R_+ consists of continuous functions g such that g' exists almost everywhere and such that $ig' + ig = f$ a.e. if $g = R_+f$. Similar statements hold for R_-.

Again let S be the closure of S_o. Then S is symmetric, so $S \pm i$ is 1-1. If we show that $(S \pm i)R_{\pm} = I$, then S is self-adjoint and $R_{\pm} = (S \pm i)^{-1}$. Let $\varphi \in C_c(\mathbb{R})$ be such that $\varphi(0) = 1$. Given $f \in C_c(\mathbb{R})$ and $g = R_+f$, let $g_n(s) = \varphi(s/n)g(s)$. Then $g_n \in D(S_o)$ and it is easy to see that $g_n \to g$, $(S_o + i)g_n \to f$. Therefore $R_+(C_c(\mathbb{R})) \subset D(S)$ and $(S+i)R_+f = f$, $f \in C_c(\mathbb{R})$. For arbitrary $f \in L^2(\mathbb{R})$, take $\{f_n\} \subset C_c(\mathbb{R})$, $f_n \to f$. Then $R_+f_n \to R_+f$ and $(S+i)R_+f_n = f_n \to f$. Thus $R_+f \in D(S)$ and $(S+i)R_+f = f$. The analogous assertions hold for R_-, so S is self-adjoint.

If T generates the translation group $\{U_t\}$ above, then we know $S \subset T$; since both are self-adjoint, $S = T$.

For later use we want to consider the operator $U = (S-i)(S+i)^{-1} = I - 2i(S+i)^{-1} = I - 2iR_+$. Then $U = u(S)$, where $u(s) = (s-i)(s+i)^{-1}$. Since $u(s)^*u(s) = 1$, U is unitary. Let $\mathcal{H}_+$ be the subspace of $\mathcal{H} = L^2(\mathbb{R})$ consisting of those $f \in \mathcal{H}$ such that $f = 0$ a.e. for $s < 0$. From (3.4) we see that $R_+(\mathcal{H}_+) \subset \mathcal{H}_+$. Therefore $U(\mathcal{H}_+) \subset \mathcal{H}_+$. We want to compute the dimension of $\mathcal{H}_+ \ominus U(\mathcal{H}_+) = \mathcal{H}_+ \cap (U(\mathcal{H}_+))^\perp$, the orthogonal complement of $U(\mathcal{H}_+)$ as a subspace of $\mathcal{H}_+$. Now $f \perp U(\mathcal{H}_+)$ if and only if $0 = (f, U_g)$ $= (U^* f, g)$, all $g \in \mathcal{H}_+$. This is true if and only if $U^* f = 0$ a.e. for $s > 0$. Now $U^* = I + 2i(S - i)^{-1} = I + 2iR_-$, so the condition is

$$f(s) = -2iR_- f(s) = 2\int_s^\infty e^{s-t} f(t)dt \quad \text{a.e.}, \quad s > 0.$$

Therefore f is continuous, hence continuously differentiable for $s > 0$ and $f' = -2f + f = -f$, $s > 0$, so $f(s) = ce^{-s}$ for some constant c (strictly speaking, of course, $f(s) = ce^{-s}$ a.e. for $s > 0$). If also $f \in \mathcal{H}_+$, then $f(s) = 0$ a.e. for $s < 0$. It follows that $\mathcal{H}_+ \ominus U(\mathcal{H}_+)$ is one-dimensional and spanned by the function $f_o(s)$ with $f_o(s) = 0$, $s \leq 0$, $f_o(s) = e^{-s}$, $s > 0$.

As a second example, let $\mathcal{H}_+$ be as above; we can consider it as $L^2(\mathbb{R}_+)$, $\mathbb{R}_+ = [0, \infty)$. Let T_o have domain consisting of all continuous f such that f' is continuous on $\mathbb{R}_+$, $f(0) = 0$, and $f(s) = 0$ for $s \geq M = M(f)$. Let $T_o f = if'$. Again, T_o is symmetric, hence has a closure T. It is easily verified that $D(T^*)$ contains all f with f' continuous and $f, f' \in \mathcal{H}_+$ (in particular, there is no condition at $s = 0$). Therefore $f(s) = e^{-s} \in D(T^*)$, and $T^* f = -if$. Thus $T^* + i$ is not 1-1, so T^* is not symmetric, $T^* \neq T$, and $T - i$ is not onto. Thus T is densely defined, closed, and symmetric, but not self-adjoint.

3. Functions of $i\frac{d}{ds}$

Since S of the preceding section is a self-adjoint operator, we can in principle define $\varphi(S)$ for a large class of functions φ. In this section (which is not necessary for subsequent chapters) we sketch the construction of $\varphi(S)$ explicitly.

Suppose φ and η are bounded continuous functions on $\mathbb{R}$ and suppose $\int|\eta(s)|\,ds < \infty$, $\int\eta(s)ds = 1$. For $\varepsilon > 0$, let $\eta_\varepsilon(s) = \varepsilon^{-1}\eta(\varepsilon^{-1}s)$. It is well-known and easily verified that the convolutions $\eta_\varepsilon * \varphi$ converge uniformly to φ; here

$$(3.6) \qquad (\eta_\varepsilon * \varphi)(s) = \int \eta_\varepsilon(s-t)\varphi(t)dt .$$

In turn, $\eta_\varepsilon * \varphi$ is approximated uniformly by taking Riemann sums, giving functions of the form

$$(3.7) \qquad \psi(s) = \Sigma\, \eta_\varepsilon(s - t_j)\varphi(t_j)(t_j - t_{j-1}) .$$

Now suppose η is a bounded rational function holomorphic on $\mathbb{R}$; then the same is true of $\eta_\varepsilon(z) = \varepsilon^{-1}\eta(\varepsilon^{-1}z)$. It follows from (3.6) and (3.7) that

$$(3.8) \qquad (\eta_\varepsilon * \varphi)(S) = \int \varphi(t)\eta_{\varepsilon,t}(S)dt ,$$

where

$$\eta_{\varepsilon,t}(s) = \eta_\varepsilon(s-t) = \varepsilon^{-1}\eta(\varepsilon^{-1}(s-t)).$$

Taking the (strong) limit in (3.8) as $\varepsilon \to 0$, we have

$$(3.9) \qquad \varphi(S) = \lim_{\varepsilon\to 0} \int \varphi(t)\eta_{\varepsilon,t}(S)dt ,$$

valid for any bounded continuous φ, and any self-adjoint operator S. In particular, let us take $\eta(s) = \pi^{-1}(1+s^2)^{-1} = (2\pi i)^{-1}[(s-i)^{-1} - (s+i)^{-1}]$.

Then $\eta_{\varepsilon,t}(s) = (2\pi i)^{-1}[(s-t-i\varepsilon)^{-1} - (s-t+i\varepsilon)^{-1}]$. Thus (4.8) becomes

$$\text{(3.10)} \qquad \varphi(S) = \lim_{\varepsilon\to 0} \int \frac{1}{2\pi i} \varphi(t)[(S-t-i\varepsilon)^{-1} - (S-t+i\varepsilon)^{-1}]dt$$

for any bounded continuous φ, the limit existing in the strong sense.

Let us compute this for S as in section 2. As in the derivation of (3.4) and (3.5), inverting $(S-t \pm i\varepsilon)$ amounts to solving a differential equation, and the solution is given by

$$\text{(3.11)} \qquad (S-t-i\varepsilon)^{-1}f(s) = i\int_s^{\infty} e^{(\varepsilon-it)(s-r)}f(r)dr$$

$$-(S-t+i\varepsilon)^{-1}f(s) = i\int_{-\infty}^{s} e^{(-\varepsilon-it)(s-r)}f(r)dr\ .$$

Suppose $f, \varphi \in C_c(\mathbb{R})$. Then we can apply (3.10) and (3.11) and interchange the integrations to get

$$\text{(3.12)} \qquad \varphi(S)f(s) = \int_{-\infty}^{\infty} f(r)[\frac{1}{2\pi}\int_{-\infty}^{\infty} e^{-i(s-r)t}\varphi(t)dt]\,dr$$

$$= \frac{1}{\sqrt{2\pi}}\int_{-\infty}^{\infty} f(r)\,\hat{\varphi}(s-r)dr\ ,$$

where $\hat{\varphi}(s) = \frac{1}{\sqrt{2\pi}}\int e^{-ist}\varphi(t)dt$ is the Fourier transform of φ.

4

VECTOR AND OPERATOR-VALUED FUNCTIONS ON THE LINE

1. Spaces of vector-valued functions

Throughout this chapter, $\mathcal{H}$ is assumed to be a separable Hilbert space, possibly finite-dimensional. By "measure" we shall mean a non-zero σ-finite positive Borel measure on $\mathbb{R}$. A function $f: \mathbb{R} \to \mathcal{H}$ is said to be measurable if for each fixed $y \in \mathcal{H}$ the scalar-valued function $s \to (f(s), y)$ is Borel measurable. If so, then also $\|f(s)\|$ is Borel measurable, since $\|f(s)\| = \sup |(f(s), y_n)|$ with the supremum taken over any countable subset $\{y_n\}$ dense in the unit ball of $\mathcal{H}$.

Let μ be a measure. Let $L^2(\mu; \mathcal{H})$ be the collection of (equivalence classes of) measurable functions $f: \mathbb{R} \to \mathcal{H}$ for which $\int \|f(s)\|^2 d\mu(s)$ is finite. If f and g are in $L^2(\mu; \mathcal{H})$, then the Schwarz inequality (1.1) and polarization identity (1.2) show that $(f(s), g(s))$ is measurable and μ-integrable. Thus $L^2(\mu; \mathcal{H})$ is an inner product space with respect to

$$(f, g) = \int (f(s), g(s)) d\mu(s) , \qquad \|f\|^2 = (f, f).$$

In fact, it is a Hilbert space. Suppose $\{f_n\}$ is a Cauchy sequence in $L^2(\mu; \mathcal{H})$. Passing to a subsequence and renumbering, we may suppose $\|f_n - f_{n+1}\| \leq 2^{-n}$. Then as in the scalar case since

$\Sigma \int \| f_n(s) - f_{n+1}(s) \| \, d\mu(s) < \infty$, we have $\{f_n(s)\}$ a Cauchy sequence in $\mathcal{H}$ for μ -- a.e. s. Let $\{y_m\}$ be a (finite or countable) orthonormal basis for $\mathcal{H}$. Then $\lim_n (f_n(s), y_m)$ is equal μ --a.e. to a Borel measurable function $\varphi_m(s)$. If we set $f(s) = \Sigma \varphi_m(s) y_m$, then f is measurable and $\|f - f_n\| = \lim \|f_n - f_{n+k}\| \leq 2^{-n}$, all n. Thus $L^2(\mu; \mathcal{H})$ is complete.

A function $f \in L^2(\mu; \mathcal{H})$ is said to be simple if it is of the form $f(s) = \Sigma k_j(s) x_j$, a finite sum with $x_j \in \mathcal{H}$ and k_j the characteristic function of a Borel set of finite μ-measure. We claim that the simple functions are dense in $L^2(\mu; \mathcal{H})$. It is easy to see that the subset of $L^2(\mu; \mathcal{H})$ consisting of those f with $f(s) = 0$ except on a set of finite measure is dense. Suppose f is such a function. For any $\varepsilon > 0$, let $\{x_n\}$ be a countable dense subset of $\mathcal{H}$ with $x_1 = 0$ and let $B_n = \{s \mid \|x(s) - x_n\| \leq \varepsilon\}$, $C_n = B_n - (B_1 \cup B_2 \cup \ldots \cup B_{n-1})$, k_n the characteristic function of C_n. Let $g_n = k_1 x_1 + \ldots + k_n x_n$ and $g = \Sigma k_j x_j$; since the C_j are mutually disjoint g has at most one non-zero summand at each point. Clearly $\|g_n(s) - g(s)\|^2 \to 0$ and $\|g_n(s) - g(s)\|^2 \leq \|g(s)\|^2$. Also, $\|g(s) - f(s)\| \leq \varepsilon$ where $f(s) \neq 0$ and $g(s) = 0$ where $f(s) = 0$. Therefore $g \in L^2(\mu; \mathcal{H})$, $\|g - f\|^2 \leq \varepsilon^2 M$ where $M = \mu(\{x \mid f(s) \neq 0\})$, and $\|g_n - g\| \to 0$. Thus f can be approximated by simple functions. Note also that any simple function can be approximated by simple functions of the form $\Sigma k_j x_j$ with x_j in a fixed countable dense subset and k_j the characteristic function of an open interval with rational endpoints; the set of such functions is countable. Let us collect the above remarks.

PROPOSITION 4.1. Let $\mathcal{H}$ be a separable Hilbert space and let μ be a positive σ-finite Borel measure on $\mathbb{R}$. The space $L^2(\mu; \mathcal{H})$ is a separable Hilbert space, and the simple functions are dense.

2. Multiplication operators

Let $\mathcal{H}$ be a separable Hilbert space and μ a measure. A measurable complex-valued function φ defines in a natural way an operator φ^X in $L^2(\mu; \mathcal{H})$. Let $D(\varphi^X) = \{f \in L^2(\mu; \mathcal{H}) | \int |\varphi|^2 \|f\|^2 d\mu < \infty\}$ and let $\varphi^X f(s) = \varphi(s)f(s)$.

PROPOSITION 4.2. The operator φ^X is closed and densely defined. The adjoint of φ^X is $(\varphi^*)^X$.

<u>Proof.</u> The domain of φ^X contains every simple function of the form $k(s)x$, $x \in \mathcal{H}$ and k the characteristic function of a Borel set on which φ is bounded. The span of such functions is dense. Suppose $f \in D((\varphi^X)^*)$ and $(\varphi^X)^* f = g$. Taking inner products with functions of the above form we find that $g(s) = \varphi^*(s)f(s)$ a.e. Therefore $(\varphi^X)^* \subset (\varphi^*)^X$; the converse inclusion is immediate. Since also $\varphi^X = ((\varphi^*)^X)^*$, φ^X is closed. ∎

In particular, if φ is real-valued, φ^X is self-adjoint, and if $|\varphi| \equiv 1$, φ^X is unitary.

Let $\mathcal{H}$ and $\mathcal{K}$ be two separable Hilbert spaces and μ, ν two measures. A bounded operator S from $L^2(\mu; \mathcal{H})$ to $L^2(\nu; \mathcal{K})$ is said to <u>commute</u> with the multiplication operator φ^X if S takes the domain of φ^X (as operator in $L^2(\mu, \mathcal{H})$) into the domain as operator in

$L^2(\nu; \mathcal{K})$ and satisfies $S(\varphi^X f) = \varphi^X Sf$. It is true, but not obvious, that if S commutes with φ^X it also commutes with $(\varphi^*)^X$. More generally, the following is true.

LEMMA. Suppose $S: L^2(\mu; \mathcal{H})$ into $L^2(\nu; \mathcal{K})$ and S commutes with φ^X. Let h be a Borel measurable function on $\mathbb{C}$. Then S commutes with $[h(\varphi)]^X$.

Proof. Let $\mathcal{C}$ be the collection of all Borel subsets $C \subset \mathbb{C}$ with the property that if $f \in L^2(\mu; \mathcal{H})$ vanishes μ-a.e. on the complement of $\varphi^{-1}(C)$, then Sf vanishes ν-a.e. there.

It is easy to see that $\mathcal{C}$ is closed under the taking of unions and countable intersections. Suppose C is the closed disc with center λ and radius ε. Then $f \in L^2(\mu; \mathcal{H})$ vanishes a.e. on the complement of $\varphi^{-1}(C)$ if and only if

$$\{ \varepsilon^{-n} \| (\varphi - \lambda)^n f \| \mid n = 1, 2, 3, \ldots \}$$

is a bounded set. If so, then the fact that S commutes with φ^X implies that (if $f \in D(\varphi^X)$), Sf satisfies the same condition, i.e. $Sf = 0$ a.e. on $\varphi^{-1}(C)'$. Thus $\mathcal{C}$ contains all closed discs, and therefore all Borel sets. Now let k be the characteristic function of a Borel set C and suppose $f \in L^2(\mu; \mathcal{H})$ is in $D(\varphi^X)$. Since $k(\varphi)^X f$ vanishes on the complement $\varphi^{-1}(C)'$, so does $S(k(\varphi)^X f)$. Since $(1 - k(\varphi))^X f$ vanishes on $\varphi^{-1}(C)$, so does $S((1 - k(\varphi))^X f)$. Thus $S(k(\varphi)^X f) = k(\varphi)^X Sf$. The assertion is true for characteristic functions; taking linear combinations and limits, it is true for all Borel functions. ∎

PROPOSITION 4.3. Let $\mathcal{H}$, $\mathcal{K}$ be separable Hilbert spaces and let μ, ν be measures. Suppose $S: L^2(\mu; \mathcal{H}) \to L^2(\nu; \mathcal{K})$ commutes with φ^X for each $\varphi \in \Phi$, a family of continuous functions separating the points of $\mathbb{R}$. Then S commutes with ψ^X for any bounded Borel-measurable function ψ.

Proof. Since S commutes with sums and products of elements of Φ, we may assume Φ to be an algebra. By the lemma we may assume $\varphi^* \in \Phi$ if $\varphi \in \Phi$. Let C_1, C_2 be disjoint compact subsets of $\mathbb{R}$. By the Stone-Weierstrass theorem there is a $\varphi \in \Phi$ with $\varphi(s) < 0$ on C_1 and $\varphi(s) > 0$ on C_2. Let $k(s) = 1$, $s < 0$, $k(s) = 0$, $s \geq 0$. Then $k(\varphi) \equiv 1$ on C_1, $k(\varphi) \equiv 0$ on C_2, and $k(\varphi)^X$ commutes with S. Therefore if $f = 0$ on the complement of C_1, $Sf = S(k(\varphi)^X f) = k(\varphi)^X Sf = 0$ on C_2. Since this is true for every compact C_2 disjoint from C_1, it follows that Sf vanishes on the complement of C_1 if f does. The argument may now be completed as in the proof of the lemma. ∎

In the next section we characterize operators commuting with all bounded measurable functions.

3. Operator-valued functions

Again, let $\mathcal{H}$ and $\mathcal{K}$ be separable. An operator-valued function $s \to S(s) \in \mathcal{B}(\mathcal{H}, \mathcal{K})$ defined on $\mathbb{R}$ is said to be measurable if for each $x \in \mathcal{H}$, $y \in \mathcal{K}$ the scalar function $(S(s)x, y)$ is Borel-measurable. Let μ and ν be measures. If $S(s)$ is measurable and f is a simple function with values in $\mathcal{H}$, then clearly $S(s)f(s)$ is measurable and $\mathcal{K}$-valued. Since simple functions are dense in $L^2(\mu; \mathcal{H})$, for any $f \in L^2(\mu; \mathcal{H})$ there is a measurable $g: \mathbb{R} \to \mathcal{K}$ such that $S(s)f(s) = g(s)$ μ-a.e. If the measure $\|S(s)\|^2\nu$ is $\leq M^2\mu$ for some constant M, then it is immediate that the ν-equivalence class of g depends only on the μ-equivalence class of f, and

$$\int \|S(s)f(s)\|^2 \, d\nu(s) \leq M^2 \int \|f(s)\|^2 d\mu(s).$$

Thus $S(s)$ defines an operator $S: L^2(\mu; \mathcal{H}) \to L^2(\nu; \mathcal{K})$ with $\|S\| \leq M$; $(Sf)(s) = S(s)f(s)$. Note that if S is such an operator then S commutes with φ^X for any bounded Borel-measurable φ. It is an important fact that the converse is true.

PROPOSITION 4.4. Let $\mathcal{H}$, $\mathcal{K}$ be separable Hilbert spaces and let μ, ν be measures. Let $S: L^2(\mu; K) \to L^2(\nu; K)$ be a bounded operator. Then S commutes with φ^X for each bounded Borel-measurable φ if and only if S is of the form $Sf(s) = S(s)f(s)$ where $S(s)$ is measurable and $\|S(s)\|^2\nu \leq M^2\mu$ for some constant M; the minimal such constant is $\|S\|$.

Proof. Let $\{x_n\}$ be an orthonormal basis for $\mathcal{H}$ and let $\mathcal{H}_o$ be the linear span of $\{x_n\}$. Let φ_o be an everywhere positive Borel-measurable function which is in $L^2(\mu)$. Suppose S commutes with all φ^X. Let $f_n(s) = \varphi_o(s)x_n$ and choose a particular measurable representative for each $Sf_n(s)$. Define $S(s)x_n$ for each n and s by $S(s)x_n = \varphi_o(s)^{-1}Sf_n(s)$. Extend $S(s)$ to $\mathcal{H}_o$ by linearity. Any simple function f with values in $\mathcal{H}_o$ can be expressed as a finite sum $\Sigma \varphi_n f_n$, where the φ_n are Borel-measurable; if f vanishes outside a set on which φ_o^{-1} is bounded, then the φ_n will be bounded. Suppose so. Then

$$Sf(s) = \Sigma \varphi_n(s)Sf_n(s) = \Sigma \varphi_n(s)\varphi_o(s)S(s)x_n = S(s)f(s)$$

ν-a.e. If φ is a simple Baire function such that $\varphi_o^{-1}\varphi$ is bounded and $f(s) = \varphi(s)x$, $x \in \mathcal{H}_o$, then

$$\int |\varphi(s)|^2 \|S(s)x\|^2 d\nu(s) = \|Sf\|^2 \leq \|S\|^2 \int |\varphi(s)|^2 d\mu(s) \|x\|^2.$$

Therefore $\|S(s)x\|^2 \nu \leq \|S\|^2 \mu$. Taking the supremum over a countable subset of $\mathcal{H}_o$ dense in the unit ball, we get $\|S(s)\|^2 \nu \leq \|S\|^2 \mu$. It follows that $S(s)$ is bounded ν-a.e., hence has a unique extension to an operator in $\mathcal{B}(\mathcal{H}, \mathcal{K})$. Redefining $S(s)$ to be zero on the exceptional set, we have $s \to S(s) \in \mathcal{B}(\mathcal{H}, \mathcal{K})$ everywhere. Moreover $(S(s)x_n, y) = (\varphi_o(s))^{-1}(Sf_n(s), y)$ ν-a.e. is measurable for each n and each $y \in \mathcal{K}$. It follows that $S(s)$ is measurable. The operator defined by $S(s)$ agrees with S on a dense subspace, hence is S.

Conversely we have already noted that $\|S(s)\|^2 \nu \leq M^2 \mu$ implies $\|S\| \leq M$. Therefore the minimal M is $\|S\|$. ∎

We want to relate some properties of $S(s)$, S, μ, and ν.

PROPOSITION 4.5. Let $S: L^2(\mu, \mathcal{H}) \to L^2(\nu; \mathcal{K})$ be determined by $S(s)$ as above. Then

(a) S is an isometry if and only if $\mu \ll \nu$ and $\psi(s)^{-1}S(s)$ is an isometry μ-a.e., where $\psi^2 = d\mu/d\nu$ is the Radon-Nikodym derivative.

(b) S is unitary if and only if $\mu \sim \nu$ and $\psi(s)^{-1}S(s)$ is unitary μ-a.e.

(c) If $\mu = \nu$ and $\mathcal{H} = \mathcal{K}$, then S is a projection if and only if $S(s)$ is a projection μ-a.e.

<u>Proof.</u> (a). Suppose S is an isometry and suppose $x \in \mathcal{H}$, φ a Borel-measurable function in $L^2(\mu)$. Let $f(s) = \varphi(s)x$. Then

$$\int |\varphi(s)|^2 \|x\|^2 d\mu = \|f\|^2 = \|Sf\|^2 = \int |\varphi(s)|^2 \|S(s)x\|^2 d\nu(s).$$

Fixing x and letting φ vary, we see that $\|x\|^2\mu = \|S(s)x\|^2 \nu \leq \|S\|^2\nu = \nu$. Taking $x \neq 0$ we have $\mu << \nu$ and $\|S(s)\| = 0$ ν-a.e. on any set of μ-measure 0. Letting x vary over a countable dense subset we get $\|S(s)x\|^2/\|x\|^2 = \psi(s)^2$ ν-a.e., where ψ^2 is the Radon-Nikodym derivative $d\mu/d\nu$. Therefore if we take $\psi(s)^{-1}S(s)$ to be zero wherever $S(s)$ is zero, we have $\psi(s)^{-1}S(s)$ an isometry a.e. Conversely if this is the case then reasoning backwards we get $\|Sf\| = \|f\|$ for any f of the above form; these generate a dense subspace, so S is an isometry.

(b) Suppose S is unitary. Then $\mu << \nu$ by (a). Taking adjoints in the equation $S\varphi^X = \varphi^X S$, we see that S^* also commutes with each φ^X. Since it is unitary, $\nu << \mu$. Thus $\mu \sim \nu$. Moreover, S^* is determined by an operator-valued function $T(s)$. We have $f(s) = S^*Sf(s) = T(s)S(s)f(s)$ a.e. for each s; taking $f_n(s) = \varphi_o(s)x_n$ with $\varphi_o > 0$, $\varphi_o \in L^2(\mu)$, and $\{x_n\}$ dense, we see that $T(s)S(s) = I$ a.e. Similarly $S(s)T(s) = I$ a.e. In particular, $S(s)$ is onto a.e. Since $\psi(s)^{-1}S(s)$ is isometric a.e., this implies it is unitary a.e. Conversely, if $\mu \sim \nu$ and $\psi(s)^{-1}S(s)$ is unitary a.e., let S be determined by $T(s)$. Then again $T(s)S(s) = I$ a.e. But $S(s)$ is invertible a.e., so this implies $S(s)T(s) = I$ a.e., which implies $SS^* = I$. Thus S is unitary.

(c) Suppose $\mathcal{H} = \mathcal{K}$, $\mu = \nu$, and suppose S is a projection. As above, $S(s) = S(s)^2 = S(s)^*$ a.e., so $S(s)$ is a projection a.e. The converse is equally easy. $\blacksquare$

5

SPECTRAL REPRESENTATION OF SELF-ADJOINT AND UNITARY OPERATORS

1. Existence of a spectral representation

Operators S and T in Hilbert spaces $\mathcal{H}$ and $\mathcal{K}$ respectively are said to be unitarily equivalent if there is a unitary operator $U: \mathcal{K} \to \mathcal{H}$ such that $T = U^* S U$; if so, we write $S \cong T$. This is clearly an equivalence relation. Moreover if $S \cong T$ and S is closed, densely defined, self-adjoint, and/or unitary, so is T.

Suppose $\mathcal{H} = \oplus \mathcal{H}_\alpha$, $\mathcal{K} = \oplus \mathcal{K}_\alpha$, and $S_\alpha : \mathcal{H}_\alpha \to \mathcal{K}_\alpha$ is linear. Define $S = \oplus S_\alpha : \mathcal{H} \to \mathcal{K}$ by letting $D(S) = \{(x_\alpha) \mid x_\alpha \in D(S_\alpha)$ and $\Sigma \|S_\alpha x_\alpha\|^2 < \infty\}$, $S(x_\alpha) = (S_\alpha x_\alpha)$. It is not hard to see that S is densely-defined, closed, self-adjoint, and/or unitary if and only if each S_α is. Moreover, if $S_\alpha \cong T_\alpha$, all α, then $S \cong T = \oplus T_\alpha$.

Let μ be a positive σ-finite Borel measure on $\mathbb{R}$ (once again we shall simply say "measure") and let $S = S(\mu)$ be φ^X in $L^2(\mu) = L^2(\mu; \mathbb{C})$, where $\varphi(s) \equiv s$. By Proposition 4.2, S is self-adjoint. More generally, if $\{\mu_\alpha\}$ is a family of measures and $S = \oplus S(\mu_\alpha)$, then S is self-adjoint. Conversely, any self-adjoint operator is unitarily equivalent to an operator of this form. Such a representation of S as a direct sum of multiplication operators is called a spectral representation.

THEOREM 5.1. Suppose S is self-adjoint. There is a family $\{\mu_\alpha\}$ of measures such that $S \cong \oplus S(\mu_\alpha)$.

Proof. Let $S: \mathcal{H} \to \mathcal{H}$. Given $x \in \mathcal{H}$, $\|x\| = 1$, let $\mathcal{H}_x$ be the closure of $\{f(S)x \mid f \in C(\tilde{\sigma}(S))\}$. For any such f, clearly $f(S)\mathcal{H}_x \subset \mathcal{H}_x$, $f(S)^*\mathcal{H}_x \subset \mathcal{H}_x$. Therefore also $f(S)\mathcal{H}_x^\perp \subset \mathcal{H}_x^\perp$. It follows that if $\|y\| = 1$, $y \in \mathcal{H}_x^\perp$, then $\mathcal{H}_y \perp \mathcal{H}_x$. By Zorn's lemma, there is a family $\{x\}$ of elements of norm 1 such that the subspaces $\mathcal{H}_x$ are mutually orthogonal and $\bigvee \mathcal{H}_x = \mathcal{H}$. Let T_x be the restriction to $\mathcal{H}_x$ of $(S - i)^{-1}$. Then in a natural way $(S - i)^{-1} \cong \oplus T_x$ and this in turn gives $S \cong \oplus S_x$, where $S_x: \mathcal{H}_x \to \mathcal{H}_x$ is self-adjoint; it is the restriction of S to $\mathcal{H}_x$. Moreover, $f(S_x)x = f(S)x$, and $\sigma(S_x) \subset \sigma(S)$ (since $(\lambda - S)^{-1}$ restricted to $\mathcal{H}_x$ is $(\lambda - S_x)^{-1}$). Therefore $\{f(S_x)x \mid f \in C(\tilde{\sigma}(S_x))\}$ is dense in $\mathcal{H}_x$. This allows us to reduce the problem to the case of a single summand: $\mathcal{H} = \mathcal{H}_x$, $\|x\| = 1$.

In this case, let $C_o(\mathbb{R})$ be the algebra of continuous functions vanishing at ∞. The map $f \to f(S)x$ is a linear mapping from $C_o(\mathbb{R})$ to a dense subspace of $\mathcal{H}$. Also

$$|(f(S)x, x)| \leq \|f(S)\| \leq \sup |f(s)|,$$

and $f \geq 0$ implies $(f(S)x, x) \geq 0$. Therefore there is a unique finite positive Borel measure μ on $\mathbb{R}$ such that $\int f(s)d\mu(s) = (f(S)x, x)$. Then

$$\int f(s)\overline{g(s)}\, d\mu(s) = ((fg^*)(S)x, x) = (f(S)x, g(S)x).$$

Therefore $f \to f(S)x$ is an isometry from a dense subspace of $L^2(\mu)$ to a dense subspace of $\mathcal{H}$. Let U be the unique unitary extension of this map. Let $S(\mu)$ be the multiplication operator in $L^2(\mu)$. Then $(S(\mu)-i)^{-1}$ is g^X where $g(s) = (s-i)^{-1}$. Therefore for $f \in C_0(\mathbb{R})$, $U((S(\mu)-i)^{-1}f) = g(S)f(S)x = (S-i)^{-1}f(S)x = (S-i)^{-1}U(f)$. This gives $U^*(S-i)^{-1}U = (S(\mu)-i)^{-1}$, so $U^*SU = S(\mu)$. $\blacksquare$

The representation of S as a direct sum of multiplication operators is not unique. For example, let μ_1, μ_2, and μ_3 be the restrictions of Lebesgue measure to the intervals $(0,1], (1,2]$, and $(0,2]$ respectively. Then in a natural way $S(\mu_3) \cong S(\mu_1) \oplus S(\mu_2)$. In the next section we consider the question of a unique representation when $\mathcal{H}$ is separable.

2. The canonical spectral representation

Let $\mu_1, \mu_2, \mu_3, \ldots, \mu_\infty$ be a collection of mutually singular non-negative σ-finite Borel measures on $\mathbb{R}$; we allow some, but not all, μ_p to vanish. Let $\mathfrak{K}_p$ be a Hilbert space of dimension p, $p = 1, 2, 3, \ldots$, and let $\mathfrak{K}_\infty$ be a separable, infinite-dimensional Hilbert space. Let $S_p(\mu_p)$ be the multiplication operator φ^X, $\varphi(s) \equiv s$, acting in $L^2(\mu_p; \mathfrak{K}_p)$, and let $S((\mu_p)) = \oplus S_p(\mu_p)$. Then $S((\mu_p))$ is a self-adjoint operator acting in a separable space. We want to show that any self-adjoint operator in a separable space is unitarily equivalent to one of this form, and that the measures μ_p are determined up to equivalence.

First, let us note that there is another, equivalent form for this model. We may assume that the spaces $\mathfrak{K}_p$ are chosen with $\mathfrak{K}_1 \subset \mathfrak{K}_2 \subset \ldots \subset \mathfrak{K}_\infty$ and $\bigvee_{p<\infty} \mathfrak{K}_p = \mathfrak{K}_\infty$. Let $\mu = \Sigma \mu_p$ and let $\{B_p\}$ be a family of disjoint Borel sets such that $\mu_p = k_p\mu$, where k is the characteristic function of B_p. Suppose $f_p \in L^2(\mu_p; \mathfrak{k}_p)$, all p, and $\Sigma \|f_p\|^2 < \infty$. Then $f = \Sigma k_p f_p \in L^2(\mu; \mathfrak{k}_\infty)$. The map $(f_p) \to f$ is easily seen to be a unitary map of $\oplus L^2(\mu_p; \mathfrak{k}_p)$ onto a subspace of $L^2(\mu; \mathfrak{k}_\infty)$. Moreover, it commutes with φ^X.

THEOREM 5.2. Let S be a self-adjoint operator in a separable Hilbert space $\mathcal{H} \neq (0)$. There is a family of mutually singular non-negative measures $\mu_1, \mu_2, \ldots, \mu_\infty$ such that $S \cong \oplus S_p(\mu_p)$. If $\mu_1', \mu_2', \ldots, \mu_\infty'$ is another such family, then $\mu_p' \sim \mu_p$, all p.

Proof. By Theorem 5.1 we may assume $S = \oplus S(\nu_j)$, the sum being at most countable (since $\mathcal{H}$ is separable). The measures ν_j may be chosen so that $\nu_1 >> \nu_2 >> \nu_3 >> \ldots$. To see this, take the decomposition $\nu_2 = \nu_2' + \nu_2''$, where $\nu_2' << \nu_1$ and $\nu_2'' \perp \nu_1$. There are corresponding Borel sets C_1, C_2', C_2'' whose complements have measure 0 with respect to ν_1, ν_2', ν_2'' respectively, and with $C_2' \subset C_1$, $C_1 \cap C_2'' = \emptyset$. By restricting functions to these sets it is easy to see that there is a natural unitary map taking $L^2(\nu_1) \oplus L^2(\nu_2) \to L^2(\nu_1) \oplus L^2(\nu_2') \oplus L^2(\nu_2'') \to L^2(\nu_1 + \nu_2'') \oplus L^2(\nu_2')$. This map takes $S(\nu_1) \oplus S(\nu_2)$ to $S(\nu_1 + \nu_2'') \oplus S(\nu_2')$. Therefore we may replace ν_1 by $\nu_1 + \nu_2''$ and ν_2 by ν_2', and assume $\nu_1 >> \nu_2$. Continuing, we may assume $\nu_1 >> \nu_3$, $\nu_1 >> \nu_4, \ldots$. Repeating the process we may then get $\nu_2 >> \nu_3$, etc. Countably many steps produce $\nu_j >> \nu_{j+1}$, all j.

Next, there are Borel sets $\mathbb{R} = B_1 \supset B_2 \supset B_3 \supset \ldots$ such that $\nu_j \sim \nu'_j$, where $\nu'_j(B) = \nu_1(B \cap B_j)$. We may even assume $\nu_j = \nu'_j$. To see this, let ψ_j be a measurable function such that $\psi_j^2 = d\nu_j/d\nu'_j$. Then $f \to f \cdot \psi_j$ is easily seen to be a unitary mapping of $L^2(\nu_j)$ onto $L^2(\nu'_j)$ which takes φ^X to φ^X. Thus we shall assume $\nu_j(B) = \nu_1(B \cap B_j)$. Let $C_p = B_p \sim B_{p+1}$ (the set-theoretic difference) and let $C_\infty = \mathbb{R} \sim (\bigcup_{p<\infty} C_p)$. Then the C_p are disjoint, and the measures μ_p with $\mu_p(B) = \nu_1(B \cap C_p)$ are mutually singular. Moreover $B_j = \bigcup_{p \geq j} C_p$, so $\nu_j = \sum_{p \geq j} \mu_p$. Again it follows in a natural way that

$$S(\nu_j) \cong \bigoplus_{p \geq j} S(\mu_p) .$$

Then

$$S = \oplus S(\nu_j) \cong \bigoplus_p \Big(\bigoplus_{j \leq p} S(\mu_p)\Big) = \bigoplus_p S_p ,$$

with

$$S_p = \bigoplus_{j \leq p} S(\mu_p) .$$

Now let $\mathfrak{K}_\infty$ be a separable, infinite-dimensional Hilbert space with orthonormal basis $\{x_n\}$. Let $\mathfrak{K}_p$ be the span of $x_1, x_2, \ldots, x_p$. For p finite, map $\bigoplus_{j \leq p} L^2(\mu_p)$ into $L^2(\mu_p; \mathfrak{K}_p)$ by $(f_j) \to \Sigma f_j(s)x_j$. This is unitary and takes S_p to $S_p(\mu_p)$. A similar construction takes S_∞ to $S_\infty(\mu_\infty)$. Thus $S \cong \oplus S_p(\mu_p)$.

Finally, suppose also that $S \cong \oplus S_p(\mu'_p)$, where the μ'_p are mutually singular measures. By the remarks preceding this theorem, we may take $\mathcal{H} = \oplus L^2(\mu_p; \mathfrak{K}_p)$ and $\mathcal{H}' = \oplus L^2(\mu'_p, \mathfrak{K}_p)$ to be subspaces of $L^2(\mu, \mathfrak{K}_\infty)$ and $L^2(\mu', \mathfrak{K}_\infty)$ with $\mu = \Sigma \mu_p$, $\mu' = \Sigma \mu'_p$. The projection of $L^2(\mu, \mathfrak{K}_\infty)$ onto $\mathcal{H}$ and the injection of $\mathcal{H}'$ into

$L^2(\mu', \mathfrak{K}_\infty)$ commute with φ^X. If we combine these with the unitary $U: \mathcal{H} \to \mathcal{H}$ such that $[\oplus S_p(\mu'_p)]U = U[\oplus S_p(\mu_p)]$, we get a map $V: L^2(\mu, \mathfrak{K}_\infty)$ into $L^2(\mu', \mathfrak{K}_\infty)$ such that V commutes with φ^X. By Proposition 4.3, V commutes with ψ^X for any bounded Borel-measurable ψ.

Let (B_p) be a family of disjoint Borel sets such that $\mu_p(B) = \mu(B \cap B_p)$, all Borel B. Let (B'_p) be a similar family for (μ'_p). Let ψ_{pq} be the characteristic function of $B_p \cap B'_q$. Then V commutes with $(\psi_{pq})^X$, which is the projection of $L^2(\mu, \mathfrak{K}_\infty)$ or $L^2(\mu', \mathfrak{K}_\infty)$ onto the subspace of functions vanishing a. e. outside $B_p \cap B'_q = A_{pq}$. Let μ_{pq} and μ'_{pq} be the restrictions of μ_p and μ'_q to A_{pq}. What we have shown, in effect, is that U takes functions vanishing outside A_{pq} to functions vanishing outside A_{pq}; thus $U: L^2(\mu_{pq}, \mathfrak{K}_p)$ into $L^2(\mu'_{pq}, \mathfrak{K}_q)$. Similarly for U^*, so U gives a unitary map U_{pq} between these spaces. Moreover, U_{pq} commutes with all ψ^X. By Proposition 4.5(c) it follows that $\mu_{pq} \sim \mu'_{pq}$ and either $\dim \mathfrak{K}_p = \dim \mathfrak{K}_q$, so $p = q$, or else $\mu_{pq} = \mu'_{pq} = 0$. Thus $\mu_p = \Sigma \mu_{pq} \sim \Sigma \mu'_{pq} = \mu'_p$. ∎

The argument used for scalar functions in the course of this proof works as well for vector functions and shows that $S_p(\mu_p) \cong S_p(\mu'_p)$ if $\mu_p \sim \mu'_p$. Thus there is a one-to-one correspondence between (unitary) equivalence classes of self-adjoint operators in separable spaces and countable families of equivalence classes of mutually singular measures. Another way of putting this is the following. Let μ be a measure on $\mathbb{R}$ and let $m(s)$ be a Borel-measurable function taking the values $0, 1, 2, \ldots, \infty$. Assume that

$m^{-1}(\{0\})$ is the largest open set with μ-measure 0. Let μ_p be the restriction of μ to $m^{-1}(\{p\})$, $p > 0$. Then the μ_p are mutually singular and $\Sigma\, \mu_p = \mu$. Conversely, if μ_p is a family of mutually singular measures, let $\mu = \Sigma\, \mu_p$. There are disjoint Borel sets B_p with $\mu_p = \mu$ restricted to B_p and the complement of $\bigcup B_p$ being the largest open set with μ-measure 0. Let $m(s) = p$, $s \in B_p$, $m(s) = 0$, $s \notin \bigcup B_p$. Then (μ, m) induce the family (μ_p). Put $(\mu, m) \cong (\mu', m')$ if $\mu \sim \mu'$ and $m = m'$ a.e. Then we have shown in effect that there is a 1-1 correspondence between equivalence classes of self-adjoint operators and equivalence classes of such pairs (μ, m). If (μ, m) corresponds to S, m is called the <u>spectral multiplicity function</u> of S.

In the finite dimensional case it is easy to see that the measure μ associated with S is a direct sum of point measures, with $\mu(\{\lambda\}) \neq 0$ if and only if $\lambda \in \sigma(S)$. Moreover $m(\lambda)$ is the dimension of the eigenspace corresponding to λ, i.e., $m(\lambda) = \dim N(S - \lambda)$. Thus in this case (the equivalence classes of) μ and S are both determined by m alone. In general, however, it is obvious that μ is not uniquely determined by m.

Let us note that if $S = \oplus\, S_p(\mu_p)$ is considered as acting in a subspace $\mathscr{H}$ of $L^2(\mu, \mathscr{K}_\infty)$, $\mu = \Sigma\, \mu_p$, then we can easily identify $\psi(S)$ for $\psi \in \mathfrak{A}(\sigma(S))$: it is just the restriction to $\mathscr{H}$ of the multiplication operator ψ^X. In particular, the projection operator E_t takes $f \in$ to g, where $g(s) = f(s)$, $s \leq t$ and $g(s) = 0$, $s > t$.

3. Unitary operators.

We shall derive the analogues for unitary operators of some of our results for self-adjoint operators. This can be done by a reduction to the self-adjoint case, in effect. What makes this possible is the correspondence between self-adjoint and unitary operators given by the Cayley transform. The idea is that the linear fractional transformation η with $\eta(z) = (z-i)/(z+i)$ maps the upper half plane onto the unit disc. In particular, it maps the real line into the unit circle: $\eta^*(s)\eta(s) = 1$ for s real. Therefore if S is self-adjoint, $U = \eta(S) = (S-i)(S+i)^{-1} = I - 2i(S+i)^{-1}$ is unitary. The transformation inverse to η is given by $w \to i(1+w)/(1-w)$, so formally one might expect to recover S by taking $i(I+U)(I-U)^{-1}$. Now in fact $I - U = 2i(S+i)^{-1}$ is 1-1 with inverse $\frac{-i}{2}(S+i)$, and $i(I+U)(I-U)^{-1}$ $= \frac{1}{2}[S+i+S-i] = S$.

Conversely, let U be a unitary operator such that $I-U$ is 1-1, and let $S = i(I+U)(I-U)^{-1} = iI + 2iU(I-U)^{-1} = -iI + 2i(I-U)^{-1}$. Then

$$(S-i) = 2iU(I-U)^{-1} = [-\tfrac{i}{2}(I-U)U^*]^{-1} = [\tfrac{i}{2}(I-U^*)]^{-1}$$

so $S-i$ is closed, so S is closed. Moreover, $D(S) = D(S-i)$ $= R(I-U^*)$, and $R(I-U^*) = N(I-U)$, so $D(S)$ is dense. Then $(S^*+i)^{-1} = [(S-i)^{-1}]^* = \frac{i}{2}(U-I) = (S+i)^{-1}$, so $S^*+i = S+i$ and $S^* = S$.

Now let U be an arbitrary unitary operator in $\mathcal{H}$ and let $\mathcal{H}_o = N(I-U)$. If $x \in \mathcal{H}_o$, then $U^*x = U^*(Ux) = x$. Therefore $\mathcal{H}_o$ is invariant under both U and U^*. It follows that $\mathcal{H}_1 = \mathcal{H}_o^{\perp}$ is invariant under both U and U^*. Let U_o, U_1 be the restrictions of

U to $\mathcal{H}_o$ and $\mathcal{H}_1$; then $U_o = I$ in $\mathcal{H}_o$ and U_1 is unitary, $I - U_1$ is 1-1. Let $S_1 = i(I+U_1)(I-U_1)^{-1}$. Then U_1 is the Cayley transform $(S_1 - i)(S_1 + i)^{-1} = \eta(S_1)$. We have $U \cong U_o \oplus U_1$, and we shall take $U = U_o \oplus U_1$. Suppose $p(\lambda) = \Sigma\, a_j U^j$ is a polynomial. Then $p(U_o) = \Sigma\, a_j U_o^j = p(\lambda)I$ and $p(U_1) = (p \circ \eta)(S_1)$. It follows from this and the results for self-adjoint operators that the mapping $p \to p(U)$ from polynomials to $\mathcal{B}(\mathcal{H})$ has an extension $f \to f(U)$, $f \in \mathfrak{A}(\Gamma)$, such that $f(U)^* = f^*(U)$, $(f(U)x, x) \geq 0$ if $x \in D(f(U))$ and $f \geq 0$ on $\Gamma = \{\lambda \mid |\lambda| = 1\}$, and $\|f(U)\| \leq \sup_{\Gamma} |f(\lambda)|$ if f is bounded. This is defined by putting $f(U) = f(1)I \oplus (f \circ \eta)(S_1)$. Using this functional calculus we may proceed as for self-adjoint operators and derive the following.

THEOREM 5.3. There is a 1-1 correspondence between unitary operators and spectral measures $\{E_t\}$ such that $E_o = 0$, $E_{2\pi} = I$, given by

$$U = \int_0^{2\pi} e^{it}\, dE_t \ , \quad E_t = \tilde{e}_t(U), \ 0 < t \leq 2\pi$$

where $\tilde{e}_t$ is defined on Γ by $\tilde{e}_t(e^{is}) = 1$, $0 < s \leq t$, $\tilde{e}_t(e^{is}) = 0$, $t < s \leq 2\pi$.

Given a family of mutually singular measures on $(0, 2\pi]$, (μ_p), let $U_p(\mu_p)$ be multiplication by e^{is} in $L^2(\mu_p; \mathcal{K}_p)$, where $\mathcal{K}_p$ is separable and p-dimensional, $p = 1, 2, \ldots, \infty$. Then $U = \oplus\, U_p(\mu_p)$ is unitary.

THEOREM 5.4. Let U be a unitary operator in a separable Hilbert space $\mathcal{H} \neq (0)$. There is a family of mutually singular non-negative measures $\mu_1, \mu_2, \ldots, \mu_\infty$ on $(0, 2\pi]$ such that $U \cong \oplus U_p(\mu_p)$. If $\mu'_1, \mu'_2, \ldots, \mu'_\infty$ is another such family, then $\mu'_p \sim \mu_p$, all p.

6

SHIFT OPERATORS AND APPLICATIONS

1. Shift operators

A <u>shift operator</u> in a Hilbert space $\mathcal{H} \neq (0)$ is a unitary operator U for which there is a subspace $\mathcal{H}_+$ such that

(6.1) $$U(\mathcal{H}_+) \subset \mathcal{H}_+ , \quad \bigwedge_{n=-\infty}^{\infty} U^n(\mathcal{H}_+) = (0) , \quad \bigvee_{n=+\infty}^{\infty} U^n(\mathcal{H}_+) = \mathcal{H} .$$

As an example, let $\mathcal{H}_o$ be a given Hilbert space and let $\ell^2(\mathcal{H}_o)$ be the collection of sequences $x = (x_n)_{n=-\infty}^{\infty}$ with $x_n \in \mathcal{H}_o$ and $\Sigma \|x_n\|^2 < \infty$. This is a Hilbert space $\mathcal{H}$ with respect to the inner product $(x, y) = \Sigma\, x_n y_n$. Let $U: \mathcal{H} \to \mathcal{H}$ be defined by $(Ux)_n = x_{n-1}$. Let $\mathcal{H}_+$ be the set of x in $\mathcal{H}$ with $x_n = 0$ for $n < 0$. Then U is unitary and $\mathcal{H}_+$ has properties (6.1) with respect to U. We shall call such a shift operator a <u>canonical</u> shift operator. The dimension of $\mathcal{H}_o$ is called the <u>multiplicity</u> of the canonical shift in $\ell^2(\mathcal{H}_o)$. Clearly two canonical shifts with the same multiplicity are unitarily equivalent. We shall prove the converse below.

As a second example, let $\mathcal{H}_o$ be a separable Hilbert space and let m be Lebesgue measure restricted to the interval $(0, 2\pi]$. Let $\mathcal{H} = L^2(m; \mathcal{H}_o)$ and let U be multiplication by e^{is}. Let $\mathcal{H}_+$ be the closed subspace generated by functions of the form $e^{ins}x$, $n \geq 0$, $x \in \mathcal{H}$. Since $\{e^{ins}\}$ spans a dense subspace of $L^2(m)$ and $\bigvee U^n(\mathcal{H}_+)$ contains everything of the form $e^{ins}x$, $x \in \mathcal{H}$, we see that $\bigvee U^n(\mathcal{H}_+) = \mathcal{H}$. We have $e^{ins}x \perp U^m(\mathcal{H}_+)$ for $m > n$, since $e^{ins}x \perp e^{ims}y$, $m > n$, so $\bigwedge U^m(\mathcal{H}_+) = \{e^{ins}x\}^{\perp} = (0)$.

Although it is convenient to have at least two examples of shift operators, there is a sense in which one is enough.

THEOREM 6.1. Let U be a shift operator. Thus U is unitarily equivalent to a canonical shift U' of uniquely determined multiplicity; the unitary equivalence can be chosen so as to take $\mathcal{H}'_+$ corresponding to U' onto $\mathcal{H}_+$ corresponding to U.

Proof. Let $\mathcal{H}^{(n)} = U^n(\mathcal{H}_+)$, $n = 0, \pm 1, \pm 2, \ldots$. From (6.1) we get $\mathcal{H}^{(n+1)} \subset \mathcal{H}^{(n)}$. Let $\mathcal{H}_n = \mathcal{H}^{(n)} \ominus \mathcal{H}^{(n+1)}$, so $\mathcal{H}^{(n)} = \mathcal{H}_n \oplus \mathcal{H}^{(n+1)}$. Since U is unitary and $U\mathcal{H}^{(n)} = \mathcal{H}^{(n+1)}$, we have $U(\mathcal{H}_n) = \mathcal{H}_{n+1}$. Also $\mathcal{H}_m \perp \mathcal{H}^{(m+1)} \supset \mathcal{H}_n$ for $n > m$, so the $\mathcal{H}_n$ are mutually orthogonal. Also $\mathcal{H}^{(m)} \cong \mathcal{H}_m \oplus \mathcal{H}^{(m+1)}$ $= \mathcal{H}_m \oplus \mathcal{H}_{m+1} \oplus \ldots \oplus \mathcal{H}_{m+k} \oplus \mathcal{H}^{(m+k+1)}$, all $k = 0, 1, 2, \ldots$. Since $\bigwedge \mathcal{H}^{(n)} = (0)$ we get $\mathcal{H}^{(m)} = \bigoplus_{n \geq m} \mathcal{H}_n$. Also $\bigvee \mathcal{H}^{(n)} = \mathcal{H}$ gives $\mathcal{H} = \oplus \mathcal{H}_n$.

Now iteration of $U(\mathcal{H}_n) = \mathcal{H}_{n+1}$ shows that $U^n(\mathcal{H}_o) = \mathcal{H}_n$, all n. Define $W: \ell^2(\mathcal{H}_o) \to \mathcal{H}$ by $W((x_n)) = \Sigma\, U^n x_n$. By what we have shown, W is unitary. Moreover if U' is the canonical shift in $\ell^2(\mathcal{H}_o)$, $W(U'x) = \Sigma\, U^n(U'x_n) = \Sigma\, U^n x_{n-1} = UWx$. Thus $U' = W^*UW$. Finally, $W(\mathcal{H}'_+) = \mathcal{H}_+$, clearly.

It remains to be shown that the multiplicity is uniquely determined, i.e. that if U and U' are the canonical shifts in $\mathcal{H} = \ell^2(\mathcal{H}_o)$ and $\mathcal{H}' = \ell^2(\mathcal{H}'_o)$ respectively, then $\dim \mathcal{H}_o = \dim \mathcal{H}'_o$.

Now if $\dim \mathcal{H} = \dim \mathcal{H}'$ is uncountable, then $\dim \mathcal{H}_o = \dim \mathcal{H} = \dim \mathcal{H}' = \dim \mathcal{H}'_o$. Therefore we can restrict attention to the case $\mathcal{H}$, $\mathcal{H}'$ separable.

Suppose $\mathcal{H}_o$ is separable and we consider the second example above. By what we have already shown, this is unitarily equivalent to the canonical shift in $\ell^2(\mathcal{H}_o)$; conversely, the latter must be unitarily equivalent to ψ^X in $L^2(m; \mathcal{H}_o)$, where $\psi(s) = e^{is}$. Thus what we want to prove is that if ψ^X in $L^2(m; \mathcal{H}_o)$ is unitarily equivalent to ψ^X in $L^2(m; \mathcal{H}'_o)$, then $\dim \mathcal{H}_o = \dim \mathcal{H}'_o$. But this is part of the assertion of Theorem 5.4. ∎

2. Some classical analysis: Fourier series, and the theorem of F. and M. Riesz.

Let us compute explicitly the unitary equivalence W between the canonical shift U in $\ell^2 = \ell^2(\mathbb{C})$ and $U' = \psi^X$ in $L^2(m; \mathbb{C})$. Let us normalize m so that $m((0, 2\pi]) = 1$. Let $e^n \in \ell^2$ have $(e^n)_m = \delta_{nm}$. Then $U^n e^o = e^n$. $W: \ell^2_+ \to L^2_+$, $W: [\ell^2_+ \ominus U(\ell^2_+)] \to L^2_+ \ominus U'(L^2_+)$. But these subspaces are one-dimensional and spanned by e^o and the function $e_o(s) \equiv 1$ respectively. Thus $We_o = \alpha e^o$ with $|\alpha| = 1$; we may assume $\alpha = 1$. Then $W^* e^o = e_o$ and $W^* e^n = W^* U^n e^o = (U')^n W^* e^o = e_n$, $e_n(s) = e^{ins}$. Thus $W^*(\Sigma \alpha_n e^n) = \Sigma \alpha_n e^{ins}$ and

$$
\begin{aligned}
(Wf)_n &= (Wf, e^n) = (f, W^* e^n) = (f, e_n) \\
&= \frac{1}{2\pi} \int_0^{2\pi} f(s) e^{-ins} \, ds .
\end{aligned}
$$

Thus W takes f to its Fourier series.

A more surprising application of the preceding results is the theorem of F. and M. Riesz.

THEOREM 6. 2. Let μ be a non-zero finite complex Borel measure on $(0, 2\pi]$ with vanishing negative Fourier-Stieltjes coefficients, i. e.,

$$\int_0^{2\pi} e^{-ins}\, d\mu(s) = 0 \quad , \qquad n < 0. \tag{6.3}$$

Then μ is equivalent to Lebesgue measure on $(0, 2\pi]$.

Proof. Let ν be the total variation of μ. Then $\mu = h(s)\nu$, where $h(s)$ is a Borel function with $0 < |h(s)| \leq 1$ ν-a. e. Let $\mathcal{H} = L^2(\nu)$ and let $U = \psi^X$, $\psi(s) = e^{is}$. Let $\mathcal{H}_+$ be the closed subspace generated by $\{U^n(h^*) \mid n \geq 0\}$. Then as above for the case $\nu = m$, we have (6.1) satisfied. Moreover $\mathcal{H}_+ \ominus U(\mathcal{H}_+)$ is spanned by h^*, so U is equivalent to a shift of multiplicity one. Theorem 6. 1 gives a unitary $W: L^2(m) \to \mathcal{H}$ such that $W\psi^X = \psi^X W$. In the notation of Theorem 5. 4, this says that $U_1(\nu) \cong U_1(m)$, so $m \sim \nu \sim \mu$. ∎

3. Another example.

Let $\mathcal{H} = L^2(\mathbb{R})$, Lebesgue measure understood. Let $U_t f(s) = f(s - t)$, $S = \lim_{s \to 0} \frac{1}{is}(U_s - I)$ and $U = (S - i)(S + i)^{-1}$. Let $\mathcal{H}_+ = L_+^2(\mathbb{R})$, the subspace of functions vanishing for $s < 0$.

PROPOSITION 5. 2. U is unitarily equivalent to the canonical shift of multiplicity 1.

Proof. As noted in the discussion of $\{U_t\}$ in Chapter 3, $U(\mathcal{H}_+) \subset \mathcal{H}_+$ and $\dim(\mathcal{H}_+ \ominus U(\mathcal{H}_+)) = 1$. It suffices to show that $\wedge U^n(\mathcal{H}_+) = (0)$ and $\vee U^n(\mathcal{H}_+) = \mathcal{H}$. Let $\mathcal{H}^{+\infty} = \wedge U^n(\mathcal{H}_+)$, $\mathcal{H}^{-\infty} = \vee U^n(\mathcal{H}_+)$. Clearly these are invariant under $U = I - 2i(S+i)^{-1}$ and under $U^{-1} = U^* = I + 2i(S-i)^{-1}$. Therefore they are invariant under $(S \pm i)^{-1}$. Let $P_\pm$ be the projection on $\mathcal{H}^{\pm\infty}$. Then $P_\pm$ commutes with $(S \pm i)^{-1}$. Therefore $P_\pm$ commutes with $f(S)$ for any f in the closed subalgebra of $C_o(\mathbb{R})$ generated by $(s \pm i)^{-1}$; but this is all of $C_o(\mathbb{R})$. By the construction of the functional calculus it follows that $P_\pm$ commute with $f(S)$ for any bounded continuous f. In particular $P_\pm$ commute with all U_t. Thus $U_t \mathcal{H}^{\pm\infty} = \mathcal{H}^{\pm\infty}$, all t. But $\mathcal{H}^{\infty} \subset \mathcal{H}_+$ and $\wedge U_t(\mathcal{H}_+) = (0)$, clearly, so $\mathcal{H}^{\infty} = (0)$. Similarly $\mathcal{H}_+ \subset \mathcal{H}^{-\infty}$ and $\vee U_t(\mathcal{H}_+) = \mathcal{H}$, so $\mathcal{H}^{-\infty} = \mathcal{H}$. ∎

4. The unilateral shift; decomposition of isometries and unitary operators

A unilateral shift in a Hilbert space $\mathcal{H}$ is an isometry V such that

(6.3) $$\bigwedge_{n \geq 0} V^n(\mathcal{H}) = (0).$$

As an example, let $\mathcal{H}_o$ be any Hilbert space and let $\mathcal{H}_+ = \ell^2_+(\mathcal{H}_o)$ be the set of sequences $x = (x_n)_{n=0}^{\infty}$ with $x_n \in \mathcal{H}_o$, $\Sigma \|x_n\|^2 < \infty$. This can be considered in a natural way as a subspace of the two sided sequence space $\ell^2(\mathcal{H}_o)$. Let $V = U_+$ be the restriction to $\mathcal{H}_+$ of the canonical shift in $\ell^2(\mathcal{H}_o)$. Then U_+ is a unilateral shift, the canonical unilateral shift in $\ell^2(\mathcal{H}_o)$. Again $\dim(\mathcal{H}_o)$ is called the multiplicity of U_t; this time the uniqueness of the multiplicity is obviou obvious, since $\dim \mathcal{H}_o = \dim(\mathcal{H}_+ \ominus U_+(\mathcal{H}_+))$. A construction like that in Theorem 6.1 proves the following

THEOREM 6.3. Any unilateral shift is unitarily equivalent to a canonical unilaterial shift of uniquely determined multiplicity.

COROLLARY 1. Let V be an isometry in a Hilbert space $\mathcal{H}$. Then there are uniquely determined complementary subspaces $\mathcal{H}_1$ and $\mathcal{H}_2$ which reduce V, such that $V|_{\mathcal{H}_1}$ is a unilateral shift and $V|_{\mathcal{H}_2}$ is unitary.

Proof. Let $\mathcal{H}_2 = \bigwedge_{n \geq 0} V^n(\mathcal{H})$ and $\mathcal{H}_1 = \mathcal{H}_2^{\perp}$. Since $V^n(\mathcal{H}) = VV^{n-1}(\mathcal{H}) = V^*V^{n+1}(\mathcal{H})$, we have $V(\mathcal{H}_2) = \mathcal{H}_2 = V^*(\mathcal{H}_2)$. Thus $\mathcal{H}_2$ reduces V. If $V_2 = V|_{\mathcal{H}_2}$, then V_2 is isometric and $V_2^* = V^*|_{\mathcal{H}_2}$ is onto, so V_2 is unitary. Let $V_1 = V|_{\mathcal{H}_1}$. Then $\bigwedge_{n \geq 0} V_1^n(\mathcal{H}_1) = \bigwedge_{n \geq 0} V^n(\mathcal{H}_1) \subset \mathcal{H}_1 \cap \mathcal{H}_2 = (0)$, so V_1 is a unilateral shift operator.

Suppose $\mathcal{H}_1'$ and $\mathcal{H}_2'$ are a second such pair of subspaces. Then $\mathcal{H}_2' = \bigwedge_{n \geq 0} V^n(\mathcal{H}_2') \subset \mathcal{H}_2$, while $\mathcal{H}_2 \cap \mathcal{H}_1' = \bigwedge_{n \geq 0} V^n(\mathcal{H}_1 \cap \mathcal{H}_1') \subset \bigwedge_{n > 0} V^n(\mathcal{H}_1') = (0)$. Therefore $\mathcal{H}_1' = \mathcal{H}_1$, $\mathcal{H}_2' = \mathcal{H}_2$. ∎

There is a somewhat similar (but non-unique) decomposition of an arbitrary unitary operator U in $\mathcal{H}$. Note that the direct sum of a family of shift operators is also a shift operator: as subspaces $\mathcal{H}_+$ one can take the sum of the corresponding spaces for the summands. It follows by a Zorn's lemma argument that there is a subspace $\mathcal{H}_1$

which reduces U and such that there is no non-zero subspace of $\mathcal{H}_2 = \mathcal{H}_1^{\perp}$ reducing $U_2 = U|_{\mathcal{H}_2}$ such that the restriction of U_2 to this subspace is a shift operator. Let us give an intrinsic characterization of operators with this property.

A unitary operator U in $\mathcal{H}$ will be called <u>singular</u> if every subspace which is invariant for U reduces U.

COROLLARY 2. Any isometry in a Hilbert space is unitarily equivalent to the direct sum of a unilateral shift, a shift, and a singular unitary operator.

<u>Proof.</u> In view of Corollary 1, we may assume that U is unitary. In view of the remarks following Corollary 1, we may assume that there is no non-zero subspace $\mathcal{H}_1$ reducing U such that $U|_{\mathcal{H}_1}$ is a shift operator. We shall show that this is equivalent to saying that U is singular. In fact, suppose $\mathcal{H}_1 \neq (0)$, $\mathcal{H}_1$ invariant under U. Let $\mathcal{H}_2 = \bigvee_{-\infty}^{\infty} U^n(\mathcal{H}_1)$ and $\mathcal{H}_3 = \bigwedge U^n(\mathcal{H}_1)$. Then $\mathcal{H}_2$ and $\mathcal{H}_3$ reduce U, so $\mathcal{H}_4 = \mathcal{H}_3 \ominus \mathcal{H}_2$ does also. Let $U_4 = U|_{\mathcal{H}_4}$ and let $\mathcal{H}_+ = \mathcal{H}_1 \ominus \mathcal{H}_2 \subset \mathcal{H}_4$. Then $\bigvee U_4^n(\mathcal{H}_+) = [\bigvee U^n(\mathcal{H}_1)] \ominus \mathcal{H}_2 = \mathcal{H}_3 \ominus \mathcal{H}_2 = \mathcal{H}_4$, while $\bigwedge U_4^n(\mathcal{H}_+) \subset [\bigwedge U^n(\mathcal{H}_1)] \ominus \mathcal{H}_2 = (0)$. Thus U_4 is a shift operator. By assumption $\mathcal{H}_4 = (0)$, or $\mathcal{H}_2 = \mathcal{H}_3$. Since $\mathcal{H}_2 \subset \mathcal{H}_1 \subset \mathcal{H}_3$ this means $\mathcal{H}_1 = \mathcal{H}_2$ reduces U. Conversely, suppose U is singular. If $\mathcal{H}_1$ reduces U and $U|_{\mathcal{H}_1}$ is a shift, then there is a subspace $\mathcal{H}_+ \subset \mathcal{H}_1$ which is invariant and with $\bigvee U^n(\mathcal{H}_+) = \mathcal{H}_1$, $\bigwedge U^n(\mathcal{H}_+) = (0)$. Since $\mathcal{H}_+$ must reduce U, $U^n(\mathcal{H}_+) \equiv \mathcal{H}_+$. Therefore $(0) = \mathcal{H}_+ = \mathcal{H}_1$. ∎

Let us remark that if $U = \oplus U_p(\mu_p)$ as in Theorem 5.4, then U restricted to some non-zero reducing subspace is a shift operator if and only if Lebesgue measure m on $(0, 2\pi]$ is absolutely continuous with respect to $\mu = \Sigma \mu_p$. Also, suppose ν is a non-zero measure with $\nu \ll m$, $\nu \not\sim m$. Then $U_1(\nu)$ is singular, but $U_\infty(m) \oplus U_1(\nu) \cong U_\infty(m)$. Thus the decomposition in Corollary 2 is not unique. We leave the proofs of these assertions as an exercise.

5. Invariant subspaces of shift operators

Throughout this section we shall let $\mathcal{H}$ be a separable Hilbert space, m = normalized Lebesgue measure on $(0, 2\pi]$, $\mathfrak{K} = L^2(\mathcal{H}) = L^2(m; \mathcal{H})$ and $\mathfrak{K}_+ = L^2_+(m; \mathcal{H})$ the closed subspace generated by $e^{ins}x$, $x \in \mathcal{H}$, $n \geq 0$. We let U be multiplication by $e^{is} = \psi(s)$ and let U_+ be the restriction of U to the invariant subspace $\mathfrak{K}_+$. We shall obtain a characterization of the subspaces of $\mathfrak{K}$ and $\mathfrak{K}_+$ invariant under U and U_+ respectively. The title of the section is justified by the fact that U and U_+ are respectively a shift and a unilateral shift of multiplicity dim $\mathcal{H}$, and that any shift or unilateral shift in a separable space can be taken in this form (up to unitary equivalence).

By Proposition 1.4, it suffices to consider the reducing and completely non-reducing subspaces.

THEOREM 5.4. (a). The subspaces of $\mathfrak{K}$ reducing U are precisely those of the form $\mathfrak{K}' = P(\mathfrak{K})$, where $Pf(s) = P(s)f(s)$ for some measurable operator-valued function such that P(s) is a projection for a.e.s.

(b) The subspaces of $\mathcal{K}$ which are invariant and completely non-reducing for U are precisely those of the form $\mathcal{K}' = Q(L^2_+(\mathcal{H}'))$ where $Qf(s) = Q(s)f(s)$ for some measurable operator-valued function with values in $\mathcal{B}(\mathcal{H}', \mathcal{H})$, $\mathcal{H}'$ a subspace of $\mathcal{H}$, and $Q(s)$ an isometry for a.e.s.

<u>Proof.</u> (a) Suppose $\mathcal{K}'$ reduces U. Let P be the projection on $\mathcal{K}'$. Then P commutes with $U = \psi^X$. By Propositions 4.3 and 4.5, P is determined by a measurable $P(s)$ with $P(s)$ a projection a.e. Conversely, if P is of this form, then it is a projection and commutes with U, so $\mathcal{K}' = P(\mathcal{K})$ reduces U.

(b) Suppose $\mathcal{K}'$ is invariant and completely non-reducing for U. Then $U' = U|_{\mathcal{K}'}$ is an isometry and $\bigwedge (U')^n(\mathcal{K}') = \bigwedge U^n(\mathcal{K}')$ must be (0), since it reduces U. Thus U' is a unilateral shift. There is an isometry $Q: L^2_+(\mathcal{H}') \to L^2(\mathcal{H}) = \mathcal{K}$ for $\mathcal{H}'$ of appropriate dimension such that the range of Q is $\mathcal{K}'$ and such that Q takes multiplication by ψ to $U = \psi^X$. Again by Propositions 4.3 and 4.5, Q is determined by a measurable operator-valued $Q(s)$ such that $Q(s)$ is an isometry for a.e. s. Then $\dim \mathcal{H}' \leq \dim \mathcal{H}$, so we may take $\mathcal{H}'$ to be a subspace of $\mathcal{H}$. Conversely if Q is of this form, Q is an isometry and $Q(L^2_+(\mathcal{H}') = \mathcal{K}'$ is invariant. Also, $\bigwedge U^n(\mathcal{K}') = (0)$ since $U|_{\mathcal{K}'}$ is a unilateral shift, so $\mathcal{K}'$ is completely non-reducing. ∎

If $\mathcal{H}'$ is a (closed) subspace of $\mathcal{H}$, then $L^2(\mathcal{H}')$ and $L^2_+(\mathcal{H}')$ can be considered in a natural way as subspaces of $L^2(\mathcal{H})$

and $L^2_+(\mathcal{H}')$ respectively. They clearly reduce U and U_+; $L^2(\mathcal{H}')$ is the special case arising in Theorem 6.2 (a) when $P(s)$ is constant a.e. and equals the projection on $\mathcal{H}'$.

THEOREM 6.5. (a) The subspaces of $\mathfrak{K}_+$ reducing U_+ are precisely those of the form $L^2_+(\mathcal{H}')$, where $\mathcal{H}'$ is a subspace of $\mathcal{H}$.

(b) The subspaces of $\mathfrak{K}_+$ which are invariant for U_+ and completely non-reducing are precisely those of the form $\mathfrak{K}' = Q(L^2_+(\mathcal{H}'))$, where $\mathcal{H}'$ is a subspace of $\mathcal{H}$ and $Qf(s) = Q(s)f(s)$ for a measurable operator-valued function such that $Q(s)$ is an isometry a.c. and such that $Q(s)x' \in \mathfrak{K}_+$, all $x' \in \mathcal{H}'$.

Proof. (a). Suppose $\mathfrak{K}'$ reduces U. Let $\mathcal{H}_n$ be the subspace of $\mathfrak{K}_+$ consisting of all $f(s) = e^{ins}x$, $x \in \mathcal{H}$, $n = 0, 1, 2, \ldots$. Then the $\mathcal{H}_n$ are mutually orthogonal and $\oplus \mathcal{H}_n = \mathfrak{K}_+$. Also, $U_+(\mathcal{H}_n) = \mathcal{H}_{n+1}$, $U_+^*(\mathcal{H}_{n+1}) = \mathcal{H}_n$. Let P_n be the projection of $\mathfrak{K}_+$ onto $\mathcal{H}_n$. Now U^n_+ is an isometry and so $U^n_+(U^*_+)^n$ is a projection, and by the above it takes $\mathcal{H}_m$ onto $\mathcal{H}_m$, $m > n$ and $\mathcal{H}_m$ to (0), $m \leq n$. Thus $I - U^n_+(U^*_+)^n = P_1 + \ldots + P_n$, and it follows that P_n can be expressed in terms of $U^m_+(U^*_+)^m$, $m \leq n$. Now if $\mathfrak{K}'$ reduces U_+, then this shows that $\mathfrak{K}'$ is invariant for P_n, all n. Therefore $\mathfrak{K}' = \oplus \mathcal{H}'_n$, $\mathcal{H}'_n = P_n(\mathfrak{K}_n)$. Now $\mathcal{H}'_n = U^n(\mathcal{H}'_o)$. Identifying $\mathcal{H}'_o$ in a natural way with a subspace $\mathcal{H}_o$ of $\mathcal{H}$, we have $\mathfrak{K}' = \oplus U^n(\mathcal{H}'_o) = L^2_+(\mathcal{H}_o)$.

(b). Suppose $\mathfrak{K}' \subset \mathfrak{K}_+$ is invariant, completely non-reducing for U_+. Then $\mathfrak{K}'$ is invariant for U as a subspace of $\mathfrak{K}$. Since $\bigwedge U^n(\mathfrak{K}') \subset \bigwedge U^n(\mathfrak{K}_+) = (0)$, it is completely non-reducing for U.

Thus the assertion follows from Theorem 6.4 (b). ∎

A little reflection suggests that the preceding two theorems do not answer all questions about the invariant subspaces for U and U_+, as they stand. The do provide an opening wedge, however.

Let us note the simplifications in Theorems 6.4 and 6.5 when the space $\mathcal{H}_o$ has dimension 1, i.e. when the functions in question are scalar-valued. In this case the subspaces of $\mathcal{H}_o$ are just (0) and $\mathcal{H}_o$, and the operator-valued functions can be considered as scalar functions. Thus in the one-dimensional case the subspaces of $\mathfrak{K}$ reducing U are just those of the form $\{f \mid f(s) = 0$ a.e. on $(0, 2\pi] \sim B\}$ for B a Borel subset of $(0, 2\pi]$. The completely non-reducing subspaces are of the form $\mathfrak{K}' = q\mathfrak{K}_+ = \{qf \mid f \in \mathfrak{K}_+\}$, where $|q(s)| = 1$ a.e. There are no non-trivial subspaces of $\mathfrak{K}_+$ reducing U_+, so any invariant subspace is completely non-reducing. It is of the form $q\mathfrak{K}_+$ where $|q(s)| = 1$ a.e. and $q\mathfrak{K}_+ \subset \mathfrak{K}_+$. Now $\mathfrak{K}_+$ here can be identified with the Hardy space H^2 of functions holomorphic on the unit disc in $\mathbb{C}$. It follows that q is (the boundary value of) an <u>inner function</u>, i.e., q is bounded and holomorphic on the disc, $\lim_{r \nearrow 1} q(re^{i\theta})$ exists and has modulus 1 for a.e. θ. Since the structure of the inner functions is precisely known, one can determine the precise structure of the lattice of invariant subspaces of U_+.

7

CONTINUOUS SHIFTS AND APPLICATIONS

1. Continuous shifts

The continuous analogue of a shift operator U, considered as the generator of the discrete group $\{U^n\}$, is a (strongly-continuous one-parameter) unitary group $\{U_t\} \subset \mathcal{B}(\mathcal{H})$ for which there is a closed subspace $\mathcal{H}_+$ such that

$$(7.1) \qquad U_t(\mathcal{H}_+) \subset \mathcal{H}_+, \; t \geq 0 \; ; \quad \bigwedge U_t(\mathcal{H}_+) = (0), \quad \bigvee U_t(\mathcal{H}_+) = \mathcal{H} .$$

As an example, let $\{\mathcal{H}_\alpha\}$ be a collection of copies of $L^2(\mathbb{R})$ with Lebesgue measure understood. Let $\mathcal{H} = \oplus \mathcal{H}_\alpha$ and let $\{U_t\}$ be defined by $[U_t f]_\alpha(s) = f_\alpha(s-t)$, for $f = (f_\alpha) \in \mathcal{H}$. Let $\mathcal{H}_+$ be the subspace consisting of f with $f_\alpha(s) = 0$ a.e. for $s < 0$, all α. We shall call any unitary group satisfying (7.1) for some subspace $\mathcal{H}_+$ a <u>continuous shift</u>, and call the example the <u>canonical continuous shift</u> of <u>multiplicity</u> equal to the cardinality of the set $\{\alpha\}$.

Unitary groups $\{U_t\}$ in $\mathcal{H}$ and $\{V_t\}$ in $\mathcal{K}$ are said to be <u>unitarily equivalent</u> if there is a unitary $W: \mathcal{K} \to \mathcal{H}$ such that $V_t = W^* U_t W$, all t.

THEOREM 7.1. Any continuous shift is unitarily equivalent to a canonical continuous shift of uniquely determined multiplicity.

Proof. Suppose $\{U_t\}$ is a continuous shift in $\mathcal{H}$, with subspace $\mathcal{H}_+$ satisfying (7.1). Let $S = \lim_{s\to 0} \frac{1}{is}(U_s - I)$ and $U = (S-i)(S+i)^{-1}$ the Cayley transform of the self-adjoint operator S. The same argument used in the proof of Proposition 6. shows that $U(\mathcal{H}_+) \subset \mathcal{H}_+$, $\bigwedge U^n(\mathcal{H}_+) = (0)$, $\bigvee U^n(\mathcal{H}_+) = \mathcal{H}$. Thus U is a shift operator. It follows that $U \cong \oplus\, U_\alpha$, where each U_α is a shift operator of multiplicity one. By Proposition 6. we may take $U_\alpha = (S_\alpha - i)(S_\alpha + i)^{-1}$, where S_α is the infinitesimal generator of the translation group $\{U_{t,\alpha}\}$ in a copy $\mathcal{H}_\alpha$ of $L^2(\mathbb{R})$. Then $S = i(I+U)(I-U)^{-1} \cong \oplus\, i(I+U_\alpha)(I-U_\alpha)^{-1} = \oplus\, S_\alpha$, so $U_t \cong \oplus\, U_{t,\alpha}$. This gives $\{U_t\}$ unitarily equivalent to a canonical continuous shift.

Any such equivalence $\{U_t\} \cong \{\oplus\, U_{t,\alpha}\}$ gives an equivalence of U with a shift operator of multiplicity equal to card$\{\alpha\}$. By Theorem 6.1, this is uniquely determined. ∎

2. The Weyl relations and von Neumann's theorem

Note the similarity of the conditions (7.1) to the conditions defining a spectral measure. In fact, let F_t be the projection of $\mathcal{H}$ on $U_t(\mathcal{H}_+)$ and let $E_t = I - F_t$. Then (7.1) becomes $E_s < E_t$ if $s \le t$, $E_t \to 0$ as $t \to -\infty$, $E_t \to I$ as $t \to +\infty$. The strong continuity of $\{U_t\}$ implies $E_t \to E_s$ as $t \to s$. Thus $\{E_t\}$ is a spectral measure. Now $U_t F_s(\mathcal{H}) = U_{t+s}(\mathcal{H}_+) = F_{t+s}(\mathcal{H})$ and
$U_t(F_s(\mathcal{H})^\perp) = U_t([U_s(\mathcal{H}_+)]^\perp) = U_t U_s(\mathcal{H}_t^\perp)) = \ldots = F_{t+s}(\mathcal{H})^\perp$.
Therefore $F_{s+t}U_t = U_t F_s$, or

$$E_{s+t} = U_t E_s U_{-t}. \tag{7.2}$$

Conversely, suppose $\{U_t\}$ is a unitary group, $\{E_t\}$ is a spectral measure, and (7.2) holds. Put $F_t = I - E_t$, and put $\mathcal{H}_+ = F_o(\mathcal{H})$. Then (7.2) implies $U_t(\mathcal{H}_+) = F_t(\mathcal{H})$, while the fact that $\{E_t\}$ is a spectral measure implies $\bigwedge F_t(\mathcal{H}_+) = (0)$, $\bigvee F_t(\mathcal{H}_+) = \mathcal{H}$, and $F_t(\mathcal{H}_+) \subset \mathcal{H}_+$ for $t \geq 0$. Thus conditions (7.1) for $\{U_t\}$ and $\mathcal{H}_+$ are equivalent to conditions (7.2) for $\{U_t\}$ and a spectral measure $\{E_t\}$. Theorem 7.1 has, therefore, an equivalent formulation in terms of groups and spectral measures. We leave the statement to the reader and proceed to another equivalent formulation.

Suppose $\{U_t\}$ and $\{E_t\}$ satisfy (7.2). Let $\{V_t\}$ be the unitary group corresponding to $\{E_t\}$: $V_t = \int e^{ist}\, dE_s$. Going back to the definition of $\int f(s)\, dE_s$, one can justify the following manipulation:

$$\begin{aligned} U_t V_s U_{-t} &= U_t\left(\int e^{isr}\, dE_r\right)U_{-t} \\ &= \int e^{isr}\, d(U_t E_r U_{-t}) = \int e^{isr}\, dE_{r+t} \\ &= e^{-ist}\int e^{is(r+t)} dE_{r+t} = e^{-ist} V_s , \end{aligned}$$

or

$$U_t V_s = e^{-ist} V_s U_t , \quad \text{all } s, t. \tag{7.3}$$

This equation, or family of equations, is called the <u>Weyl relations</u> for the groups $\{U_t\}$ and $\{V_t\}$.

Conversely, suppose $\{U_t\}$ and $\{V_t\}$ are two unitary groups satisfying the Weyl relations. Let $\{E_t\}$ be the spectral measures associated with $\{V_t\}$. For fixed t, the groups $\{U_t V_s U_{-t}\}$ and $\{e^{-ist} V_s\}$ correspond to the sprectral measures $\{U_t E_s U_{-t}\}$ and $\{E_{s+t}\}$ respectively. Therefore (7.3) implies (7.2). Since (7.2) leads back to (7.1), we can apply Theorem 7.1 to conclude that two groups satisfying the Weyl relations are (simultaneously) unitarily equivalent to a direct sum of copies of the corresponding groups in $L^2(\mathbb{R})$. What are these groups?

We know that $\{U_t\}$ is the translation group in this case. The subspace $U_t(\mathcal{H}_+)$ consists of those functions vanishing for $s < t$, so its orthogonal complement, $E_t(\mathcal{H})$, consists of functions vanishing for $s > t$. Therefore E_t is multiplication by the characteristic function of the interval $(-\infty, t]$. Therefore, from the construction, $\int f(s)dE_s$ is multiplication by f. In particular, V_t is multiplication by e^{ist}. An easy computation verifies that the translation group $\{U_t\}$ and multiplication group $\{V_t\}$ do satisfy the Weyl relations.

A representation of the Weyl relations is a pair of unitary groups $\{U_t\}$ and $\{V_t\}$ satisfying (7.3). Two such representations $\{U_t\}, \{V_t\}$ in $\mathcal{H}$ and $\{U_t'\}, \{V_t'\}$ in $\mathcal{H}'$ are said to be unitarily equivalent if there is a unitary $W: \mathcal{H}' \to \mathcal{H}$ with $U_t' \equiv W^* U_t W$, $V_t' \equiv W^* V_t W$. A Schrödinger representation of the Weyl relations is a direct sum of copies of the representation in $L^2(\mathbb{R})$ for which $U_t f(s) = f(s-t)$, $V_t f(s) = e^{it} f(s)$. The multiplicity of a Schrödinger representation is the cardinal number of copies in this direct sum. The following theorem is due to von Neumann. The discussion of this section shows that it is equivalent to Theorem 7.1.

THEOREM 7.2. Any representation of the Weyl relations is unitarily equivalent to a Schrödinger representation of uniquely determined multiplicity.

Let $\{U_t\}, \{V_t\}$ be a representation of the Weyl relations, and let $U_t = \exp(itS)$, $V_t = \exp(itT)$. A formal differentiation of (7.3) with respect to t at $t = 0$ gives

$$-iV_s S + iSV_s = -isV_s \,.$$

A second formal differentiation, with respect to s at $s = 0$ gives

$$TS - ST = -i \,. \tag{7.4}$$

This is the "canonical commutation relation" or Heisenberg relation of quantum mechanics; since S and T are not bounded operators, it cannot be taken literally. For the Schrödinger representation of multiplicity one, T is multiplication by s and S is the closure of $i\frac{d}{ds}$, as in Chapter 3, §3. If $f \in C_c(\mathbb{R})$ and f' is continuous, then $f \in D(ST) \cap D(TS)$ and certainly

$$s(if'(s)) - i(sf(s))' = -if(s) \,.$$

The Weyl relations can be derived formally from (7.4) with $U_t = \exp(itS)$, $V_t = \exp(itT)$. Thus if the Weyl relations are taken as the proper, well-defined form of (7.4), then Theorem 7.2 establishes the uniqueness of the representations of the Heisenberg relation. In quantum mechanics, this is the basis for asserting the equivalence of the Heisenberg and Schrödinger formulations.

3. The Plancherel theorem

The Weyl relations (7.3) are symmetric in $\{U_t\}, \{V_t\}$, except for a change of sign in e^{-ist}. Therefore if we take $\{V_{-t}\}$ and $\{U_t\}$ for $\{U_t\}$ and $\{V_t\}$ respectively, (7.3) still holds. This observation is the basis for an abstract version of the Plancherel theorem for $L^2(\mathbb{R})$.

THEOREM 7.3. Let $\{U_t\}, \{V_t\}$ be a representation of the Weyl relations in $\mathcal{H}$. There is a unitary $W: \mathcal{H} \to \mathcal{H}$ such that

$$U_t \equiv W^* V_{-t} W \ , \quad V_t \equiv W^* U_t W \ . \tag{7.5}$$

<u>Proof.</u> We may take $\{U_t\}, \{V_t\}$ to be the Schrödinger representation of multiplicity one, since the general result will follow by taking direct sums and applying Theorem 7.2. Now as we have noted, $\{V_{-t}\}, \{U_t\}$ is also a representation of the Weyl relations; all we need to show is that it has multiplicity 1. Let a be the multiplicity. We have $\{U_t, V_t\} \to \{V_{-t}, U_t\} \cong a$ copies of $\{U_t, V_t\}$. Interchanging once more, $\{V_{-t}, U_t\} \to \{U_{-t}, V_{-t}\} \cong a$ copies of $\{V_{-t}, U_t\} \cong a^2$ copies of $\{U_t, V_t\}$. Obvious $\{U_{-t}, V_{-t}\} \cong$ one copy of $\{U_t, V_t\}$; by uniqueness of multiplicity, $a^2 = 1$, and $a = 1$. ∎

Let us examine this in more detail for the Schrödinger representation of multiplicity one. Let us write $\hat{f}$ for Wf, $f \in L^2(\mathbb{R})$, and set $\hat{U}_t = V_{-t}$, $\hat{V}_t = U_t$. Then (7.5) becomes

$$(U_t f)^\wedge(s) = \hat{U}_t \hat{f}(s) = e^{-ist}\hat{f}(s) \tag{7.6}$$

$$(V_t f)^\wedge(s) = \hat{V}_t \hat{f}(s) = \hat{f}(s-t) \ . \tag{7.7}$$

These tell us what happens to translations and certain multiplications, so it is natural to calculate what W does to smooth combinations of translations and multiplications. Recall that the convolution of two functions $f, g \in C_c(\mathbb{R})$ is the function $f * g = g * f$ in $C_c(\mathbb{R})$ with

$$(f * g)(s) = \int f(s-t)g(t)dt = \int g(t)(U_t f)(s)dt .$$

Approximating by Riemann sums and applying (7.6), (7.7) gives

$$(f * g)^{\wedge}(s) = \int g(t)(U_t f)^{\wedge}(s)dt = \tilde{g}(s)\hat{f}(s) \quad \text{a.e.},$$

where

$$\tilde{g}(s) = \int e^{-ist} g(t)dt .$$

Since $f * g = g * f$,

(7.8) $$\tilde{f}(s)\hat{g}(s) = \tilde{g}(s)\hat{f}(s) \quad \text{a.e.}$$

Take $g \in C_c(\mathbb{R})$ with $\int g(t)dt = 1$ and $g(s) = 0$, $|s| \geq 1$. Let $g_n(s) = ng(ns)$. Then $\int g_n = 1$ and $g_n(s) = 0$, $|s| \geq \frac{1}{n}$. Therefore $\tilde{g}_n(s) \to 1$ uniformly on bounded sets. In particular, for each fixed s, $\tilde{g}_n(s) \neq 0$ for large n. Thus (7.8) gives

$$\hat{f}(s) = \lim_{n\to\infty} \frac{\hat{g}_n(s)}{\tilde{g}_n(s)} \tilde{f}(s) \quad \text{a.e.}, \quad f \in C_c(\mathbb{R}),$$

or

$$\hat{f}(s) = h(s)\tilde{f}(s) \quad \text{a.e.}, \quad f \in C_c(\mathbb{R}) ,$$

with h independent of f.

A simple computation gives $(V_t f)^{\sim} = \hat{V}_t \tilde{f} = U_t \tilde{f}$. Thus $hU_t\tilde{f} = h(V_t f)^{\sim} = (V_t f)^{\wedge} = U_t\hat{f} = U_t(h\tilde{f}) = (U_t h)(U_t\tilde{f})$, all $f \in C_c(\mathbb{R})$. Therefore $U_t h = h$ a.e. This implies that there is a constant α with $h(s) = \alpha$ a.e. Thus

(7.9) $$\hat{f}(s) = \alpha \int e^{-ist} f(t)dt \quad \text{a.e.}, \quad f \in C_c(\mathbb{R}) .$$

A similar argument shows that the inverse mapping $W^*: f \to \check{f}$ is given by

(7.10) $$\check{f}(s) = \beta\int e^{ist} f(t)dt \quad \text{a.e.,} \quad f \in C_c(\mathbb{R}),$$

where β is constant. The formulas (7.9) and (7.10), since they define bounded operators on dense subspaces of $L^2(\mathbb{R})$, extend to all $f \in L^1(\mathbb{R}) \cap L^2(\mathbb{R})$. Moreover if $f \in L^2(\mathbb{R})$ then the restrictions f_N to the intervals $(-N, N)$ are in $L^1(\mathbb{R})$ and converge to f in the norm of $L^2(\mathbb{R})$. Therefore the corresponding integrals in (7.9) and (7.10) converge in mean to $\hat{f}$ and $\check{f}$ respectively. Thus we have determined W and W^*: they are just the Fourier transform and inverse Fourier transform, and Theorem 7.3 is essentially the Plancherel theorem.

Explicit calculation of $\hat{f}$ for $f(s) = \exp(-s^2/2)$ shows that, since W is unitary, $|\alpha| = (2\pi)^{-1/2}$ and $\beta = \bar{\alpha}$. We may normalize by taking $\alpha = \beta = (2\pi)^{-1/2}$.

4. Invariant subspaces

In this section, let $\mathcal{H}$ be a separable Hilbert space and let $\mathcal{K} = L^2(\mathcal{H})$ with respect to Lebesgue measure on $\mathbb{R}$. The Fourier transform may be defined for this space as for scalar functions:

$$\hat{f}(s) = \frac{1}{\sqrt{2\pi}} \lim_{N\to\infty} \int_{-N}^{N} e^{-ist} f(t)dt,$$

$$\check{f}(s) = \frac{1}{\sqrt{2\pi}} \lim_{N\to\infty} \int_{-N}^{N} e^{ist} f(t)dt .$$

Here the limits exists in the norm of $\mathcal{K}$. Then $W: f \to \hat{f}$ is unitary, and equations (7.6), (7.7) carry over: these assertions are easily proved by reduction to the scalar case. Changing notation, let $U_t f(s) = e^{-ist} f(s)$. Let $L^2_+(\mathcal{H}) = \mathcal{K}_+$ be the subspace consisting of all functions of the form f, where $f(s) = 0$ a.e. for $s < 0$. Then (7.6) shows that $\mathfrak{K}_+$ is invariant under the positive semigroup $\{U_t\}_{t\geq 0}$.

In general we say that a closed subspace $\mathfrak{K}' \subset \mathfrak{K}$ is <u>invariant</u> for $\{U_t\}_{t\geq 0}$ if it is invariant for each U_t, $t \geq 0$. We say that $\mathfrak{K}'$ <u>reduces</u> $\{U_t\}_{t\geq 0}$ if it reduces each U_t, $t \geq 0$. There is an analogue of Proposition 1.4, proved in the same way: any invariant subspace for $\{U_t\}_{t\geq 0}$ is uniquely the direct sum of a reducing subspace and one which is completely non-reducing (has no non-zero reducing subspace). We have the exact analogues of Theorems 6.4 and 6.5.

THEOREM 7.4. (a). The subspaces of $\mathfrak{K}$ reducing $\{U_t\}_{t\geq 0}$ are precisely those of the form $\mathfrak{K}' = P(\mathfrak{K})$, where $Pf(s) = P(s)f(s)$ and $P(s)$ is a measurable operator-valued function such that $P(s)$ is a projection a.e.

(b) The subspaces of $\mathfrak{K}$ which are invariant and completely non-reducing for $\{U_t\}_{t\geq 0}$ are precisely those of the form $\mathfrak{K}' = Q(L^2_+(\mathcal{H}'))$, where $\mathcal{H}'$ is a subspace of $\mathcal{H}$ and $Qf(s) = Q(s)$ $Qf(s) = Q(s)f(s)$, $Q(s)$ a measurable operator-valued function such that $Q(s)$ is an isometry a.e.

Let U_t^+, $t \geq 0$, be the restriction of U_t to $\mathfrak{K}_+$. Again if $\mathcal{H}' \subset \mathcal{H}$, $L^2(\mathcal{H}')$ and $L^2_+(\mathcal{H}')$ can be considered in a natural way as subspaces of $\mathfrak{K}$ and $\mathfrak{K}_+$.

THEOREM 7.5. (a). The subspaces of $\mathfrak{K}_+$ reducing $\{U_t^+\}_{t\geq 0}$ are precisely those of the form $L_+^2(\mathcal{H}')$, $\mathcal{H}'$ a subspace of $\mathcal{H}$.

(b). The subspaces of $\mathfrak{K}_+$ which are invariant and completely non-reducing for $\{U_t^+\}_{t\geq 0}$ are precisely those of the form $\mathfrak{K}' = Q(L_+^2(\mathcal{H}'))$, where $Qf(s) = Q(s)f(s)$, $Q(s)$ a measurable operator operator-valued function such that $Q(s)$ is an isometry a.e. and $Q(s)x' \in \mathfrak{K}_+$, all $x' \in \mathcal{H}' \subset \mathcal{H}$.

Proofs can be given for both parts of Theorem 7.4 and (b) of Theorem 7.5 which exactly parallel the corresponding parts of Theorem 6.4 and Theorem 6.5. An alternative approach, which also applies to (a) of Theorem 7.5, is the following. Let S generate $\{U_t\}$ and let $U = (S - i)(S + i)^{-1} = I - 2i(S + i)^{-1}$. Formula (3.2) and the fact that $\mathfrak{K}_+$ is invariant for U_t, $t \geq 0$, shows that $\mathfrak{K}_+$ is invariant for $(S + i)^{-1}$, hence for U. Let $U^+ = U|_{\mathfrak{K}_+}$. Similarly, if $\mathfrak{K}'$ is invariant for U_t, $t \geq 0$, then it is invariant for U. Conversely, suppose $\mathfrak{K}'$ is invariant for U. Then it is also invariant for U_t, $t > 0$. In fact, let $g_\varepsilon(z) = \exp(it(z + i\varepsilon))$, $\varepsilon > 0$. Then $g_\varepsilon(z)$ is holomorphic on the upper half-plane, and $g_\varepsilon(S) = e^{-\varepsilon t}U_t \to U_t$ as $\varepsilon \to 0$. Let $\eta(z) = (z - i)(z + i)^{-1}$. Then η maps the upper half-plane to the unit disc. Expressing $g_\varepsilon \circ \eta^{-1}$ as a power series, expanded about the origin, leads to an expansion for g_ε in terms of powers of $\eta(z)$. It follows that $g_\varepsilon(S) = \Sigma\, \alpha_n U^n$, and $\mathfrak{K}'$ is invariant under $g_\varepsilon(S)$, all ε, hence under U_t, $t > 0$.

Suppose $\mathfrak{K}'$ reduces $\{U_t\}_{t\geq 0}$. Then it is invariant under U_t, $t \leq 0$, hence under $(S - i)^{-1}$, hence under U^* so it reduces U. Conversely if $\mathfrak{K}'$ reduces U, the argument just given can be modified to show that $\mathfrak{K}'$ reduces $\{U_t\}$. Finally, if $\mathfrak{K}' \subset \mathfrak{K}_+$, let P be the projection of $\mathfrak{K}$ on $\mathfrak{K}'$. Since $\mathfrak{K}_+$ is invariant under U_t, $t \geq 0$, it follows that $PU_t^+ = U_t^+P$, $t \geq 0$ if and only if $PU_t = U_tP$, $t \geq 0$. Thus $\mathfrak{K}'$ reduces $\{U_t^+\}_{t \geq 0}$ if and only if it reduces $\{U_t\}$. Similarly, $\mathfrak{K}'$ reduces U^+ if and only if it reduces U.

Now U is unitarily equivalent to the Cayley transform of the generator of the translation group. Therefore, by Proposition 6.1, U is a shift operator. This and the preceding remarks show how Theorems 7.4 and 7.5 can be reduced to Theorems 6.4 and 6.5. We leave the details to the reader.

8

POSITIVE MEASURES AND HARMONIC FUNCTIONS

1. Operator-valued measures on the line

Let $\mathcal{H}$ be a Hilbert space. An <u>operator-valued measure</u> on $\mathbb{R}$ is a function $\sigma \mapsto F(\sigma)$ from Borel sets in $\mathbb{R}$ to $\mathcal{B}(\mathcal{H})$ such that

(8. 1) $\quad F(\sigma) \geq 0$,

(8. 2) $\quad F(\bigcup \sigma_n) = \Sigma\, F(\sigma_n)$ if the σ_n are disjoint.

The sum here is taken to be the strong limit of the partial sums. These conditions are equivalent to the condition that the set function $\mu_x(\sigma) = (F(\sigma)x, x)$ is a positive finite Borel measure on $\mathbb{R}$ for each $x \in \mathcal{H}$. In fact if $F(\cdot)$ is an operator-valued measure, clearly μ_x is a positive finite Borel measure. Conversely, suppose this is the case for each $x \in \mathcal{H}$. Then for any σ and x, $(F(\sigma)x, x) = \mu_x(\sigma) \geq 0$, so $F(\sigma) \geq 0$. If $\{\sigma_n\}$ is a sequence of disjoint Borel sets and $\tau_n = \bigcup_{m \leq n} \sigma_m$, then

$$(F(\tau_n)x, x) = \mu_x(\tau_n) = \sum_{m \leq n} \mu_x(\sigma_m) = \sum_{m \leq n} (F(\sigma_m)x, x) ,$$

all x. Therefore $F(\tau_n) = \sum_{m \leq n} F(\sigma_m)$, and $F(\tau_n)$ is a non-decreasing sequence of positive operators bounded by $F(\tau)$, where $\tau = \bigcup \sigma_n$. It converges strongly to something, and by the countable additivity of μ_x the limit is $F(\tau)$.

Given an operator-valued measure $F(\cdot)$, let $F_t = F((-\infty, t])$. Then

(8.1)' $\quad 0 \le F_s \le F_t$ if $s \le t$, $F_t \le F_\infty \in \mathcal{B}(\mathcal{H})$,

(8.2)' $\quad F_t \to F_s$ if $t \searrow s$, $F_t \to 0$ if $t \searrow -\infty$.

In fact suppose $t_n \searrow s$. $F_{t_n} - F_s = F((s, t_n])$. Since $\bigcap_n (s, t_n] = \emptyset$ and $(s, t_n]$ is a decreasing sequence of sets, $(F((s,t_n])x, x) = \mu_x((s,t_n]) \to 0$, all x. Thus $F_{t_n} \to F_s$. Conversely, suppose $\{F_t\}$ is a family of operators in $\mathcal{B}(\mathcal{H})$ satisfying (8.1)' and (8.2)'. For each $x \in \mathcal{H}$, $(F_t x, x)$ is a non-negative bounded non-decreasing function on $\mathbb{R}$ which is continuous from the right and vanishes at $-\infty$. Thus there is a unique positive finite Borel measure μ_x such that $(F_t x, x) = \mu_x((-\infty, t])$. Let

$$\mu_{x,y} = \frac{1}{4}[\mu_{x+y} - \mu_{x-y} + i\mu_{x+iy} - i\mu_{x-iy}] .$$

Then $\mu_{x,y}((-\infty, t])$ is linear in x and conjugate linear in y for each t. Moreover $\mu_{x,y}$ is of finite total variation; in fact

$$\begin{aligned} |\mu_{x,y}(\sigma)| &\le \frac{1}{4}\|F(\mathbb{R})\|(\|x+y\|^2 + \|x-y\|^2 + \|x+iy\|^2 + \|x-iy\|^2) \\ &= \|F(\mathbb{R})\|(\|x\|^2 + \|y\|^2) . \end{aligned}$$

It follows that for each σ in the σ-ring generated by the $(-\infty, t]$, i.e. for each Borel set σ, there is a unique $F(\sigma) \in \mathcal{B}(\mathcal{H})$ such that $\mu_{x,y}(\sigma) = (F(\sigma)x, y)$. Now $(F(\sigma)x, x) = \mu_{x,x}(\sigma) = \mu_x(\sigma)$, so $F(\cdot)$ is an operator-valued measure and $F_t = F((-\infty, t])$.

Because of the 1-1 correspondence between operator-valued measures and operator-valued functions $\{F_t\}$ satisfying (8.1)' and

(8.2)', we can and shall abuse the terminology by referring to $\{F_t\}$ also as an operator-valued measure.

We say that $F(\cdot)$ is <u>normalized</u> if $F(\mathbb{R}) = I$, or equivalently if $F_t \to I$ as $t \to \infty$. In particular a spectral measure $\{F_t\}$ is a normalized operator-valued measure. More generally, suppose $\{E_t\}$ is a spectral measure in $\mathcal{H}$ and suppose $\mathfrak{K}$ is a closed subspace of $\mathcal{H}$. $P: \mathcal{H} \to \mathfrak{K}$ be the projection. Let F_t be PE_t restricted to $\mathfrak{K}$, considered as an operator in $\mathfrak{K}$. It is easy to see that $\{F_t\}$ is a normalized operator-valued measure in $\mathfrak{K}$; we shall say that it is the <u>compression</u> of $\{E_t\}$ to $\mathfrak{K}$, and that $\{E_t\}$ is a <u>dilation</u> of $\{F_t\}$ to $\mathcal{H}$. It is not so obvious that given a normalized $\{F_t\}$ in $\mathfrak{K}$, one can find an $\mathcal{H} \subset \mathfrak{K}$ and a spectral measure $\{E_t\}$ in $\mathcal{H}$ whose compression is $\{F_t\}$.

THEOREM 8.1. Let $\{F_t\}$ be a normalized operator-valued measure in $\mathfrak{K}$. There is a Hilbert space $\mathcal{H} \subset \mathfrak{K}$ and a spectral measure $\{E_t\}$ in $\mathcal{H}$ such that $\{F_t\}$ is the compression of $\{E_t\}$ to $\mathfrak{K}$.

<u>Proof.</u> Let $\mathcal{H}_0$ be the space of functions from $\mathbb{R}$ to which are of the form

$$f(t) = \Sigma\, k_\sigma(t) x_\sigma \qquad \text{(finite sum)},$$

where the k_σ are characteristic functions of Borel sets $\sigma \subset \mathbb{R}$. If $g = \Sigma\, k_\tau x_\tau$, let

$$<f, g> = \Sigma\, (F(\sigma \cap \tau) x_\sigma, y_\tau).$$

The finite additivity of $F(\cdot)$ implies that this is independent of the particular representations of f and g as sums. The positivity of F gives $<f,f> \geq 0$. Let $\mathcal{J} = \{f \in \mathcal{H}_o \mid <f,f> = 0\}$. Then $\mathcal{J}$ is a subspace and we define a positive definite inner product of $\mathcal{H}_o/\mathcal{J}$ by

$$(f + \mathcal{J}\,,\ g + \mathcal{J}\,) = <f, g> .$$

Let $\mathcal{H}$ be the completion of $\mathcal{H}_o/\mathcal{J}$ with respect to this inner product. We can identify $\mathcal{k}$ with a subspace of $\mathcal{H}$ by letting $Jx = f_x + \mathcal{J}$, where $f_x(t) \equiv x$. Then $(Jx, Jy) = <f_x, f_y> = (F(\mathbb{R})x, y) = (x, y)$, so J is unitary.

Let E_t^o be the operator in $\mathcal{H}_o$ of multiplication by the characteristic function of $(-\infty, t]$. Then it is easy to check that $(E_t^o)^2 = E_t^o$, $<E_t^o f, g> = <f, E_t^o g>$, and $0 \leq <E_t^o f, f> \leq <f, f>$. It follows that E_t^o induces a unique projection E_t in $\mathcal{H}$ such that $E_t(f + \mathcal{J}\,) = E_t^o f + \mathcal{J}$. Then $\{E_t\}$ is a spectral measure. Let us just check that $E_t \to E_s$ as $t \searrow s$. It suffices to note that for $f = \Sigma h_\sigma x_\sigma$ in $\mathcal{H}_o$, and with $\sigma_t = (-\infty, t]$,

$$\begin{aligned}(E_t(f + \mathcal{J}\,), f + \mathcal{J}\,) &= <E_t^o f, f> \\ &= \Sigma\,(F(\sigma \cap \tau \cap \sigma_t)x_\sigma, x_\tau) \\ \to \Sigma(F(\sigma \cap \tau \cap \sigma_s)x_\sigma, x_\tau) &= (E_s(f + \mathcal{J}\,), f + \mathcal{J}\,).\end{aligned}$$

Finally, for $x, y \in \mathcal{k}$,

$$\begin{aligned}(E_t Jx, Jy) = <E_t^o f_x, f_y> &= (F((-\infty, t])x, y) \\ &= (F_t x, y) .\end{aligned}$$

Thus $\{F_t\}$ is unitarily equivalent to the compression of $\{E_t\}$ to $J(\mathcal{k}\,) \subset \mathcal{H}$. ∎

Let $\mathcal{H}_1$ be the completion of the space $\mathcal{H}_o$ above with respect to the norm $\|f\| = \sup \|f(t)\|$. This contains, in particular, all continuous bounded functions from $\mathbb{R}$ to $\mathcal{R}$. For $f \in \mathcal{H}_o$, $\langle f, f\rangle \leq \|f\|^2$. Therefore the form $\langle f, g\rangle$ can be extended to a form on $\mathcal{H}_1$, and we can consider $\mathcal{H}$ as the completion of $\mathcal{H}_1/\mathcal{J}_1$, where $\mathcal{J}_1 = \{f \in \mathcal{H}_1 \mid \langle f, f\rangle = 0\}$.

2. Harmonic functions

An operator valued function $\lambda \longmapsto S_\lambda$ from an open set $\Omega \subset \mathbb{C}$ to $\mathcal{B}(\mathcal{H})$ is said to be <u>harmonic</u> if it is continuous in norm and if $(S_\lambda x, y)$ is a harmonic function of λ for each $x, y \in \mathcal{H}$. By polarization, it suffices to assume this when $x = y$. (The continuity condition can be shown to be superfluous.)

The real and imaginary parts of an operator $S \in \mathcal{B}(\mathcal{H})$ are defined by

$$\operatorname{Re} S = \frac{1}{2}(S + S^*), \quad \operatorname{Im} S = \frac{1}{2i}(S - S^*) .$$

These are self-adjoint and $S = \operatorname{Re} S + i \operatorname{Im} S$. Since the real and imaginary parts of a scalar-valued holomorphic function are harmonic, the same is true for operator-valued holomorphic functions. In fact

$$\operatorname{Re}(S_\lambda x, x) = (\operatorname{Re} S_\lambda x, x) , \quad x \in \mathcal{H} .$$

As in the scalar case, there is a converse.

PROPOSITION 8.1. Suppose Ω is a connected, simply connected open set in $\mathbb{C}$ and suppose S_λ is harmonic on Ω. If $S_\lambda = S_\lambda^*$ in Ω, then there is a holomorphic function T_λ on Ω with $S_\lambda \equiv \operatorname{Re} T_\lambda$. T_λ is unique up to addition of a constant function.

<u>Proof.</u> Suppose $\lambda_o \in \Omega$. By a translation and dilation we may take $\lambda_o = 0$ and assume $\{|\lambda| \leq 1\}$ is contained in Ω. Define T_λ for $|\lambda| < 1$ by

$$T_\lambda = \int_\Gamma \left(\frac{\mu+\lambda}{\mu-\lambda}\right) S_\mu \, dm(\mu) ,$$

where $\Gamma = \{|\lambda| = 1\}$ and $dm(\mu) = (2\pi i\mu)^{-1} d\mu$. Since S_μ is continuous on Γ this exists as a Riemann integral, and differentiation under the integral sign shows that T_λ is holomorphic. Since $S_\mu^* = S_\mu$ and m is a positive measure on Γ,

$$\operatorname{Re} T_\lambda = \int_\Gamma \operatorname{Re}\left(\frac{\mu+\lambda}{\mu-\lambda}\right) S_\mu \, dm(\mu) .$$

But this together with the classical Poisson integral formula for scalar functions shows that $(\operatorname{Re} T_\lambda x, y) = (S_\lambda x, y)$, $|\lambda| < 1$. Thus $\operatorname{Re} T_\lambda = S_\lambda$. Suppose T_λ' is holomorphic and $\operatorname{Re} T_\lambda' = S_\lambda$ for $|\lambda| < 1$. Then for each $x \in \mathcal{H}$, the holomorphic function $(T_\lambda x, x) - (T_\lambda' x, x)$ is purely imaginary for $|\lambda| < 1$. Therefore it is constant, and $T_\lambda' \equiv T_\lambda + (T_o' - T_o)$.

It follows that for each $\lambda_o \in \Omega$, there is an appropriate T_λ defined near λ_o. The usual analytic continuation argument and the monodromy theorem give the existence of a T_λ on all of Ω. ∎

Let $\{F_t\}$ be a normalized operator-valued measure on $\mathbb{R}$. If φ is a bounded continuous function on $\mathbb{R}$, we can define a bounded operator by $\int \varphi(t) dF_t$. In fact, let $M = \sup |\varphi(t)|$ and let $\{\sigma_j\}$ be a finite partition of $\mathbb{R}$ into Borel sets, with $t_j \in \sigma_j$. Then

$$|(\Sigma\, \varphi(t_j)F(\sigma_j)x, x)| \leq M\Sigma\,(F(\sigma_j)x, x)$$
$$= M(F(\mathbb{R})x, x) = M\|x\|^2 .$$

It follows, by polarization, that

(8.3) $$\|\Sigma\, \varphi(t_j)F(\sigma_j)\| \leq 2M .$$

This is the basis for proceding, as in the construction of the Riemann integral, to show that the limit of these sums over finer partitions exists (in norm). We denote it by $\int \varphi(t)dF_t$.

In particular, suppose $\lambda \notin \mathbb{R}$; define

$$T_\lambda = \int (t - \lambda)^{-1} dF_t .$$

It is easily seen that T_λ is holomorphic for $\lambda \notin \mathbb{R}$. For $\lambda \in \mathbb{C}_+ = \{\lambda \in \mathbb{C} \mid \operatorname{Im} \lambda > 0\}$,

$$S_\lambda = \operatorname{Im} T_\lambda = \int \operatorname{Im}(t - \lambda)^{-1} dF_t$$

is harmonic and ≥ 0 (since $\operatorname{Im}(t - \lambda)^{-1} \geq 0$).

As $s \to +\infty$, $s \operatorname{Im}(t - is)^{-1} = s^2(t^2 + s^2)^{-1}$ is bounded by 1 and converges to 1 uniformly for t in compact subsets of $\mathbb{R}$. From this and the fact that $F((-N, N)) \to I$ as $N \to \infty$ it is not hard to see that $sS_{is} \to I$ as $s \to +\infty$. Thus we have proved the first half of the following.

THEOREM 8.2. There is a 1-1 correspondence between normalized operator-valued measures $\{F_t\}$ on $\mathbb{R}$ and harmonic functions S_λ on $\mathbb{C}_+$ such that

$$S_\lambda \geq 0 \;, \quad sS_{is} \to I \quad \text{as } s \to +\infty \;,$$

given by

$$S_\lambda = \int \operatorname{Im}(t - \lambda)^{-1} dF_t .$$

Before completing the proof, let us recall the analogous result for scalar-valued harmonic functions on the unit disc: if $f(\lambda)$ is harmonic and non-negative for $|\lambda| < 1$, then

$$f(\lambda) = \int_\Gamma \operatorname{Re}\left(\frac{\mu+\lambda}{\mu-\lambda}\right) d\nu(\mu),$$

where ν is a positive finite measure on Γ. Moreover if $\sigma \subset \Gamma$ is a Borel set,

$$\nu(\sigma) = \lim_{r \nearrow 1} \frac{1}{2\pi} \int_\sigma f(r\mu)\, dm(\mu).$$

Suppose f is continuous on the closed disc except possibly at 1, and $f(\lambda) = 0$, $|\lambda| = 1$, $\lambda \neq 1$. Then the measure ν is concentrated at the point $\lambda = 1$, so $f(\lambda) = c\operatorname{Re}(1+\lambda)(1-\lambda)^{-1}$, where $c \geq 0$. There is a corresponding result for the upper half-plane $\mathbb{C}_+$ which can be obtained from this via the mapping $\lambda \to (\lambda - i)(\lambda + i)^{-1}$ taking $\mathbb{C}_+$ onto the disc.

LEMMA. If $g(\lambda)$ is harmonic and non-negative on $\mathbb{C}_+$, continuous on $\mathbb{C}_+ \cup \mathbb{R}$, and vanishes on $\mathbb{R}$, then g is of the form $g(\lambda) = c \operatorname{Im} \lambda$ for some $c \geq 0$.

Proof of Theorem 8.2. We could appeal to the result for the scalar case, but let us instead simply mimic the proof for that case. Suppose first that S_λ is harmonic and non-negative on $\mathbb{C}_+$ and continuous on $\mathbb{C}_+ \cup \mathbb{R}$. Let $\{\varphi_n\}$ be a family of continuous real-valued functions on $\mathbb{R}$ with

$$0 \leq \varphi_n(t) \leq \varphi_{n+1}(t), \quad \varphi_n(t) = 1, \quad |t| \leq n,$$

$$\varphi_n(t) = 0, \quad |t| \geq n+1 .$$

Define $F_t^{(n)}$ by $F_t^{(n)} = \frac{1}{\pi}\int_{-\infty}^{t} \varphi_n(s)S_s ds$. Since $\varphi_n \geq 0$, $S_s \geq 0$, it is easily seen that each $\{F_t^{(n)}\}$ is an operator-valued measure. Let $S_\lambda^{(n)}$ be the corresponding harmonic function:

$$(8.4) \qquad S_\lambda^{(n)} = \int \operatorname{Im}(t - \lambda)^{-1} dF_t^{(n)} = \frac{1}{\pi}\int \operatorname{Im}(t - \lambda)^{-1} \varphi_n(t) S_t \, dt.$$

This is harmonic and non-negative on $\mathbb{C}_+$. We claim it is continuous on $\mathbb{C}_+ \cup \mathbb{R}$ and equal to $\varphi_n S$ on $\mathbb{R}$. In fact if $\lambda = s + i\varepsilon$, $\operatorname{Im}(t - \lambda)^{-1} = \varepsilon((t - s)^2 + \varepsilon^2)^{-1} = \psi_\varepsilon(t - s) = \varepsilon^{-1}\psi(\varepsilon^{-1}(t - s))$, where $\psi(t) = (t^2 + 1)^{-1}$. Thus

$$S_{s+i\varepsilon}^{(n)} = \frac{1}{\pi}\int \psi_\varepsilon(t - s)\varphi_n(t) S_t dt.$$

Now $\varphi_n S$ is continuous on $\mathbb{R}$, $\frac{1}{\pi}\int \psi_\varepsilon = 1$, and $\int_{|\lambda| \geq \delta} \psi_\varepsilon(t)dt \to 0$ as $\varepsilon \to 0$ for each $\delta \to 0$. The standard argument involving approximate identities shows, therefore, that $S_{s+i\varepsilon}^{(n)} \to \varphi_n(s)S_s$ as $\varepsilon \searrow 0$.

This implies $S_s^{(n)} \leq S_s$, $s \in \mathbb{R}$. Moreover (8.4) and the fact that $\varphi_n(t) = 0$, $|t| \geq n+1$ imply that $\|S_\lambda^{(n)}\| \to 0$ as $|\lambda| \to \infty$, $\lambda \in \mathbb{C}_+ \cup \mathbb{R}$. Thus for any $\varepsilon > 0$, there is an N such that $S_\lambda^{(n)} \leq S_\lambda + \varepsilon I$, $|\lambda| \geq N$, $\lambda \in \mathbb{C}_+$. It follows, by applying the maximum principle to the scalar functions $(S_\lambda^{(n)}x, x)$ and $(S_\lambda x, x) + \varepsilon(x, x)$, that $S_\lambda^{(n)} \leq S_\lambda + \varepsilon I$ for all $\lambda \in \mathbb{C}_+$. Thus $S_\lambda^{(n)} \leq S_\lambda$, $\lambda \in \mathbb{C}_+$. Now

$$\frac{1}{\pi}\int_{-n}^{n} S_t dt \leq F^{(n)}(\mathbb{R}) \quad \lim_{s \to +\infty} sS_{is}^{(n)}$$

$$\leq \lim_{s \to +\infty} sS_{is} = I, \quad \text{all } n.$$

Thus $\frac{1}{\pi}\int S_t dt \leq I$. Then we may define $F_t^{(\infty)}$ by

$F_t^{(\infty)} = \frac{1}{\pi}\int_{-\infty}^{t} V_s\,ds$ and $V_\lambda^{(\infty)}$ by

$$V_\lambda^{(\infty)} = \int \operatorname{Im}(t-\lambda)^{-1}dF_t^{(\infty)} = \frac{1}{\pi}\int \operatorname{Im}(t-\lambda)^{-1}V_t\,dt .$$

The above argument shows $V_{s+i\varepsilon}^{(\infty)} \to V_s$ as $\varepsilon \searrow 0$, $s \in \mathbb{R}$. Moreover clearly $V_\lambda^{(n)} \to V_\lambda^{(\infty)}$ as $n \to \infty$, so $V_\lambda^{(\infty)} \le V_\lambda$. Given $x \in \mathcal{H}$, apply the lemma to $g(\lambda) = (V_\lambda x, x) - (V_\lambda^{(\infty)}x, x)$. Since $g(is) \to 0$ as $s \to \infty$, $G(\lambda) \equiv 0$. Thus $V_\lambda = V_\lambda^{(\infty)}$ has the desired form.

Now suppose V_λ is not assumed continuous on $\mathbb{C}_+ \cup \mathbb{R}$. Define $V_\lambda^{(\varepsilon)} = V_{\lambda+i\varepsilon}$, $\varepsilon > 0$. This is continuous on $\mathbb{C}_+ \cup \mathbb{R}$, harmonic, positive, and satisfies

$$sV_{is}^{(\varepsilon)} = sV_{is+i\varepsilon} \sim (s+\varepsilon)V_{i(s+\varepsilon)} \to I \quad \text{as} \quad s \to +\infty.$$

Thus

$$V_\lambda^{(\varepsilon)} = \int \operatorname{Im}(t-\lambda)^{-1}dF_t^{(\varepsilon)} .$$

Let X be the set of normalized positive Borel measures on $\mathbb{R}$. This is compact with respect to the weak* topology as a subset of the dual space of the space of continuous functions on $\mathbb{R}$ vanishing at ∞. For each $x \in \mathcal{H}$ with $\|x\| = 1$, let X_x be a copy of X, and let Y be the Cartesian product $\prod X_x$. This is compact with respect to the product topology. Let $\mu_x^{(\varepsilon)} \in X_x$ be the measure $(F^{(\varepsilon)}(\sigma)x, x)$, $\varepsilon > 0$. The net $(\mu_x^{(\varepsilon)}) \in Y$ has a subnet converging as $\varepsilon \to 0$ to an element $(\mu_x) \in Y$. This means that for each σ the corresponding subnet of the net of operators $\{F^{(\varepsilon)}(\sigma)\}$ converges weakly to an operator $F(\sigma)$ such that $(F(\sigma)x, x) = \mu_x(\sigma)$, $\|x\| = 1$. Then $F(\cdot)$ is an operator-valued measure. Let $S_\lambda^{(0)} = \int \operatorname{Im}(t-\lambda)^{-1}dF_t$. Then for each x with $\|x\| = 1$,

$$(S_\lambda^{(0)}x, x) = \int \operatorname{Im}(t - \lambda)^{-1} d\mu_x(t) = \lim_{\varepsilon} \int \operatorname{Im}(t - \lambda)^{-1} d\mu_x^{(\varepsilon)}(t)$$

$$= \lim (S_\lambda^{(\varepsilon)}x, x) = (S_\lambda x, x) .$$

Thus $S_\lambda = S_\lambda^{(0)}$ has a representation of the desired form.

Finally, we want to show that this is unique. Let $\{F_t'\}$ be another operator-valued measure with $S_\lambda = \int \operatorname{Im}(t - \lambda)^{-1} dF_t'$. Let $\mu_x'(\sigma) = (F'(\sigma)x, x)$. With ψ_ε as above

$$\frac{1}{\pi} (S_{s+i} x, x) = \int \psi_\varepsilon (s - t) d\mu_x'(t).$$

Thus $\frac{1}{\pi} \int_\sigma (V_{s+i\varepsilon} x, x) ds = (\psi_\varepsilon * \mu_x')(\sigma)$. The measure $\psi_\varepsilon * \mu_x'$ converges in the weak * topology to μ_x' as $\varepsilon \to 0$, so μ_x' is uniquely determined by S_λ. $\blacksquare$

Let us take note of the analogous result for the lower half plane. Suppose T_λ is harmonic and $T_\lambda \leq 0$ for $\operatorname{Im}\lambda < 0$. Let $S_\lambda = -T_{\overline{\lambda}}$. This is ≥ 0 and harmonic in $\mathbb{C}_+$. If $-sT_{-is} \to I$ then $sS_{is} \to I$ as $s \to +\infty$. In this case

$$T_\lambda = -S_{\overline{\lambda}} = -\int \operatorname{Im}(t - \overline{\lambda})^{-1} dF_t$$

$$= \int \operatorname{Im}(t - \lambda)^{-1} dF_t$$

for a (unique) normalized operator-valued measure.

9

DISSIPATIVE OPERATORS

1. Maximal dissipative operators

An operator $T: \mathcal{H} \to \mathcal{H}$ is said to be <u>dissipative</u> if $\operatorname{Im}(Tx, x) \geq 0$ for all $x \in D(T)$. In particular if S is symmetric, both S and $-S$ are dissipative (and conversely). A dissipative operator T is said to be <u>maximal dissipative</u> if $T \subset S$, S dissipative implies $S = T$. Let $\mathbb{C}_- = \{\lambda \mid \operatorname{Im} \lambda < 0\}$.

PROPOSITION 9.1. Suppose T is densely defined, closed, and dissipative. The following are equivalent.

(a) T is maximal dissipative.

(b) $T^* - \bar{\lambda}$ is 1-1 for some $\lambda \in \mathbb{C}_-$.

(c) $T - \lambda$ is invertible for some $\lambda \in \mathbb{C}_-$.

(d) $T - \lambda$ is invertible for all $\lambda \in \mathbb{C}_-$.

<u>Proof.</u> Suppose $\lambda = \mu - i\nu \in \mathbb{C}_+$, $\mu, \nu \in \mathbb{R}$, $\nu > 0$. For $x \in D(T)$,

$$\|(T - \lambda)x\| \, \|x\| \geq \operatorname{Im}((T - \lambda)x, x) \geq \nu \|x\|^2 ,$$

so $\|(T - \lambda)x\| \geq \nu \|x\|$. Therefore $T - \lambda$ is 1-1. Since it is closed and we have this estimate, it is easy to see that the range $R(T - \lambda)$ is closed, and that the inverse is a bounded map from $R(T - \lambda)$ to $\mathcal{H}$. Thus $(T - \lambda)$ is invertible (in $\mathcal{H}$) if and only if $R(T - \lambda) = \mathcal{H}$, or $(0) = R(T - \lambda)^{\perp} = N(T^* - \bar{\lambda})$. This shows that (b) $\Longrightarrow$ (c).

Clearly (d) $\Rightarrow$ (c). Moreover, (c) $\Rightarrow$ (a). In fact suppose $T \subset S$, S dissipative. Then $S - \lambda$ is 1-1, $S-\lambda \supset T-\lambda$, and $T-\lambda$ is onto. It follows that $S-\lambda = T-\lambda$, so $S = T$.

To complete the proof it suffices to show that (a) implies $T^*-\bar{\lambda}$ is 1-1 for <u>all</u> $\lambda \in \mathbb{C}_-$. Suppose $\lambda \in \mathbb{C}_-$, and suppose $x \in N(T^*-\bar{\lambda}) \cap D(T)$. Then

$$0 \leq \operatorname{Im}(Tx, x) = \operatorname{Im}(x, T^*x) = \operatorname{Im}(x, \bar{\lambda}x)$$
$$= (\operatorname{Im} \lambda)\|x\|^2 \leq 0,$$

so $x = 0$. Thus $N(T^*-\bar{\lambda}) \cap D(T) = (0)$ and we can define an operator S on $D(T) + N(T^* - \bar{\lambda})$ by

$$Sx = Tx, \quad x \in D(T), \qquad Sy = \bar{\lambda}y, \quad y \in N(T^* - \bar{\lambda}).$$

Then $S \supset T$ and is a proper extension if and only if $T^*-\bar{\lambda}$ is not 1-1. Suppose $x \in D(T)$, $y \in N(T^*-\bar{\lambda})$. Then

$$\begin{aligned}(S(x+y), (x+y)) &= (Tx, x) + (Tx, y) + \bar{\lambda}(y, x) + \bar{\lambda}(y, y)\\ &= (Tx, x) + (x, T^*y) + \bar{\lambda}(y, x) + \bar{\lambda}(y, y)\\ &= (Tx, x) + 2\operatorname{Re}(x, \bar{\lambda}y) + \bar{\lambda}(y, y).\end{aligned}$$

Therefore

$$\operatorname{Im}(S(x+y), x+y) \geq \operatorname{Im}\bar{\lambda}\,\|y\|^2 \geq 0 . \tag{9.1}$$

We have shown that T has a proper dissipative extension if $T^*-\bar{\lambda}$ is not 1-1. ∎

PROPOSITION 9.2. A densely defined, closed dissipative operator has a maximal dissipative extension.

<u>Proof.</u> Let T be densely defined, closed, and dissipative. Let $\lambda \in \mathbb{C}_-$ and let S be defined as in the proof of Proposition 9.1.

Then S is densely defined. We claim it is closed and maximal dissipative. Suppose $\{x_n\} \subset D(T)$, $\{y_n\} \subset N(T^* - \bar{\lambda})$ and $x_n + y_n \to$, $S(x_n + y_n) \to$. Applying (9.1) with $x = x_n - x_m$, $y = y_n - y_m$, we see that $\{y_n\}$ converges. Say $y_n \to y$; then $y \in N(T^* - \bar{\lambda})$ since $T^* - \bar{\lambda}$ is closed. Since also $x_n + y_n \to$, we have $x_n \to x$. Also, $S(x_n + y_n) = Tx_n + \bar{\lambda} y_n \to$, so $Tx_n \to$. Since T is closed this gives $x \in D(T)$, $Tx_n \to Tx$. Thus $x + y \in D(S)$, $S(x_n + y_n) \to S(x + y)$.

To show S maximal dissipative, it suffices to show $S^* - \bar{\lambda}$ 1-1. But $S^* - \bar{\lambda} \subset T^* - \bar{\lambda}$, so $N(S^* - \bar{\lambda}) \subset N(T^* - \bar{\lambda}) \subset D(S)$. Since S is dissipative, as was shown in the preceding proof, we must have $N(S^* - \bar{\lambda}) \cap D(S) = (0)$. Thus $N(S^* - \bar{\lambda}) = (0)$. $\blacksquare$

If T is maximal dissipative, $(T - \lambda)^{-1}$ is defined and holomorphic for $\lambda \in \mathbb{C}_-$. We can use the results of Chapter 8 to give an integral representation for $(T - \lambda)^{-1}$.

THEOREM 9.1. Suppose T is closed, densely defined, and maximal dissipative. Then there is a unique normalized operator-valued measure $\{F_t\}$ on $\mathbb{R}$ such that

$$(9.2) \qquad (\lambda - T)^{-1} = \int (\lambda - t)^{-1} dF_t, \quad \lambda \in \mathbb{C}_- .$$

<u>Proof.</u> Let $T_\lambda = \text{Im}(T - \lambda)^{-1}$. This is harmonic on $\mathbb{C}_-$. If $x \in \mathcal{H}$ and $y \; \lambda \; (T - \lambda)^{-1}x$, then $\text{Im}(y, x) = \text{Im}(y, (T-\lambda)y) = -\text{Im}(Ty, y) + \text{Im}\,\lambda \, \|y\|^2 \le 0$. Thus $T_\lambda \le 0$, $\lambda \in \mathbb{C}_-$. Note that $\|(T + is)^{-1}\| \le s^{-1}$, $s > 0$, so $\|is(T + is)^{-1}\| \le 1$. For $x \in D(T)$,

$$\begin{aligned} is(T + is)^{-1}x &= (T + is)^{-1}[is + T - T]x \\ &= x - (T + is)^{-1}(Tx) \to x \quad \text{as } s \to +\infty. \end{aligned}$$

Thus $is(T + is)^{-1} \to I$ and $-sT_{-is} \to I$ as $s \to +\infty$. It follows that there is a unique $\{F_t\}$ with

$$-T_\lambda = \int \operatorname{Im}(\lambda - t)^{-1} dF_t \quad , \quad \lambda \in \mathbb{C}_- .$$

$$= \operatorname{Im} \int (\lambda - t)^{-1} dF_t .$$

Thus the holomorphic functions $(\lambda - T)^{-1}$ and $\int (\lambda - t)^{-1} dF_t$ have the same imaginary part. They differ by a constant. But both $\to 0$ as $\operatorname{Im} \lambda \to -\infty$, so they coincide. ∎

2. Self-adjoint dilations

Let T be a (closed, densely defined) maximal dissipative operator in $\mathfrak{K}$. A self-adjoint operator S in a Hilbert space $\mathcal{H} \supset \mathfrak{K}$ is said to be a dilation of T if

(9.3) $$((\lambda - T)^{-1}x, y) = ((\lambda - S)^{-1}x, y),$$

all $\lambda \in \mathbb{C}_-$, $x, y \in \mathfrak{K}$. This is the same as saying that $(\lambda - T)^{-1}$ is the restriction to $\mathfrak{K}$ of $P(\lambda - S)^{-1}$, where P is the projection of $\mathcal{H}$ on $\mathfrak{K}$. The dilation S is said to be minimal if there is no proper subspace of $\mathcal{H}$ which reduces S and contains $\mathfrak{K}$. If there were such a subspace $\mathcal{H}_1$, then $S_1 = S|_{\mathcal{H}_1}$ would also be a dilation of T. Theorem 8.1 has an analogue for dissipative operators.

THEOREM 9.2. A closed densely defined maximal dissipative operator T has a minimal self-adjoint dilation S. Moreover S is unique up to unitary equivalence.

Proof. Let $\{F_t\}$ be such that $(\lambda - T)^{-1} = \int (\lambda - t)^{-1} dF_t$. Let $\mathcal{H}$ be the space constructed in the proof of Theorem 8.1, and let

$\mathcal{H}_1$, $\mathcal{J}_1$ be as in the remarks following that proof. Then $\mathcal{H}_1$ is closed, and $\mathcal{J}_1$ invariant, under multiplication by the scalar function $(\lambda - t)^{-1}$ for $\lambda \notin \mathbb{R}$. This gives rise to an operator S_λ in $\mathcal{H}$ with $\|S_\lambda\| \leq \sup_{\mathbb{R}} |(\lambda - t)^{-1}|$, $S_\lambda^* = S_{\overline{\lambda}}$, and $S_\lambda - S_\mu = (\mu - \lambda)^{-1} S_\lambda S_\mu$. Moreover since $|\lambda - t|^{-2} \geq \delta > 0$ on $\mathbb{R}$, S_λ is easily seen to be 1-1. The expressions for $S_\lambda - S_\mu$ shows that $S = -\lambda - S_\lambda^{-1}$ is independent of $\lambda \notin \mathbb{R}$. Since $S_\lambda^* = S_{\overline{\lambda}}$, S is self-adjoint. Moreover $S_\lambda = (\lambda - S)^{-1}$. For $x, y \in \mathfrak{K}$ it is not hard to see that

$$((\lambda - S)^{-1}(f_x + \mathcal{J}), f_y + \mathcal{J}_1) = \int (\lambda - t)^{-1} d(F_t x, y)$$
$$= ((\lambda - T)^{-1} x, y) .$$

Thus S is a dilation of $JTJ^* \cong T$. We claim that S is minimal. In fact any closed subspace of $\mathcal{H}$ reducing S and containing $Jx = f_x + \mathcal{J}_1$ for all $x \in \mathfrak{K}$ must also contain $\varphi(t) f_x(t) + \mathcal{J}_1$ for any φ which is a uniform limit of linear combinations of the $(\lambda - t)^{-1}$. These comprise all continuous functions on $\mathbb{R}$ which vanish at ∞. Moreover, taking an appropriate decreasing sequence of such functions we find that the subspace contains everything of the form $k_\sigma f_x + \mathcal{J}$, where $\sigma = [a, b]$ is a closed interval. It follows that it contains $\mathcal{H}_1 / \mathcal{J}_1$, which is dense in $\mathcal{H}$. Thus S is minimal.

Now suppose S_1, S_2 are self-adjoint operators in $\mathcal{H}_1, \mathcal{H}_2$ respectively. Suppose $J_j : \mathfrak{K} \to \mathcal{H}_j$ is unitary and

(9.4) $$((\lambda - S_j)^{-1} J_j x, J_j y) = ((\lambda - T)^{-1} x, y),$$

$\lambda \in \mathbb{C}_-$, $x, y \in \mathfrak{K}$. Finally, suppose $\bigvee_{\lambda \notin \mathbb{R}} (\lambda - S_j)^{-1} J_j(\mathfrak{K}) = \mathcal{H}_j$, $j = 1, 2$. We want to show that $S_1 \cong S_2$. Let $U_o = J_2 J_1^*$ be the unitary map from $J_1(\mathfrak{k})$ to $J_2(\mathfrak{k})$. Then (9.4) gives

$$((\lambda - S_2)^{-1}U_o x, U_o y) = ((\lambda - S_1)^{-1}x, y), \tag{9.5}$$

$x, y \in J_1(\mathfrak{k})$, $\lambda \in \mathbb{C}_-$. Taking adjoints, this is also true for $\lambda \in \mathbb{C}_+$. Linear combinations of the functions $\varphi(t) = (\lambda - t)^{-1}$, $\lambda \notin \mathbb{R}$ are uniformly dense in the algebra of all continuous functions on $\mathbb{R}$ which vanish at ∞. Therefore for any such f, g we have

$$\begin{aligned}(f(S_2)U_o x, g(S_2)U_o y) &= (g^*(S_2)f(S_2)U_o x, U_o y) \\ &= (g^*(S_1)f(S_1)x, y) = (f(S_1)x, f(S_1)y),\end{aligned} \tag{9.6}$$

$x, y \in J_1(\mathfrak{k})$. Thus U_o extends to a unitary map U from the closed subspace of $\mathcal{H}_1$ generated by all $f(S_1)x$ onto the closed subspace of $\mathcal{H}_2$ generated by all $f(S_2)U_o x$. By minimality these subspaces are all of $\mathcal{H}_1$ and $\mathcal{H}_2$ respectively. Moreover from the form of U, clearly $(\lambda - S_1)^{-1} = U^*(\lambda - S_2)^{-1}U$, $\lambda \notin \mathbb{R}$. Thus $S_1 \cong S_2$. ∎

Either Theorem 9.1 or Theorem 9.2 can be made the basis for a functional calculus for maximal dissipative operators. Let $r_\lambda(t) = (\lambda - t)^{-1}$, where $\lambda \in \mathbb{C}_-$. Then (9.2) and (9.3) give

$$r_\lambda(T) = \int r_\lambda(t)dF_t \tag{9.7}$$

$$(r_\lambda(T)x, y) = (r_\lambda(S)x, y), \tag{9.8}$$

where S is a self-adjoint dilation of T. Since we already have a functional calculus for S, we can extend it to T using (9.8). On the other hand, starting from (9.7) we could develop a functional calculus directly which would give another approach to the self-adjoint case. Let us sketch the latter approach. Let $\mathcal{F}_+$ be the algebra of functions on $\mathbb{C}$ generated by r_λ, $\lambda \in \mathbb{C}_-$, and by $e(\lambda) \equiv \lambda$. Define $\varphi(T)$ for $\varphi \in \mathcal{F}_+$ by

$$(9.9) \qquad \varphi(T) = \int \varphi(t) dF_t .$$

Since these functions are bounded on $\mathbb{R}$, the integral makes sense. It is clearly a linear map from $\mathcal{F}_+$ to $\mathcal{B}(\mathcal{H})$, and we claim that it is multiplicative. Now for $\lambda \neq \mu$, $r_\lambda r_\mu = (\lambda - \mu)^{-1}[r_\mu - r_\lambda]$, so $(r_\lambda r_\mu)(T) = (\lambda - \mu)^{-1}(r_\mu(T) - r_\lambda(T) = r_\lambda(T) r_\mu(T)$ by the resolvent equation. But functions of the form $\alpha e + \Sigma \alpha_j r_{\lambda_j}$ with λ_j's distinct are dense in $\mathcal{F}_+$ with respect to the norm $\|f\| = \sup_{\mathbb{R}} |f(t)|$, and the inequality (8.3) implies

$$(9.10) \qquad \|\varphi(T)\| \leq 2\|\varphi\| .$$

Therefore $\varphi \mapsto \varphi(T)$ is multiplicative. (Note that if T is self-adjoint then (9.7) holds for all $\lambda \notin \mathbb{R}$, so we could work with the algebra $\mathcal{F}$ instead of $\mathcal{F}_+$ and recover the results of Chapter 2.) Moreover we can extend this algebra homomorphism to the closed subalgebra of the algebra of bounded continuous functions on $\mathbb{R}$ which is generated by $\mathcal{F}_+$. Moreover the inequality (9.10) can be improved. In fact if we let S be as constructed in the proof of Theorem 9.2 and let $\varphi(S)$ be the operator in $\mathcal{H}$ generated by multiplication by φ in $\mathcal{H}_1$, then (independent of the functional calculus for general self-adjoint operators) we can show $\|\varphi(S)\| \leq \|\varphi\|$. But also $(\varphi(S)x, y) = (\varphi(T)x, y)$, $x, y \in \mathcal{K}$. Therefore

$$(9.10)' \qquad \|\varphi(T)\| \leq \|\varphi\| .$$

The map $\varphi \longmapsto \varphi(T)$ can be extended to a still larger algebra of functions, but we shall mention only a particular family of functions. Let $\varphi_{s,n}(t) = (1 - ist/n)^{-n}$ for $s \geq 0$, $n = 1, 2, \ldots$, and let $\varphi_s(t) = \exp(ist)$. Then $\|\varphi_{s,n}\| = 1$, $\varphi_{s,n} \to \varphi$ uniformly on compact subsets of $\mathbb{R}$. From this and the fact that $F((-m, m))x \to x$ as $m \to \infty$ for each $x \in \mathfrak{k}$ we get that $\varphi_{s,n}(T) \to V_s$, where

$$V_s = \int \exp(ist)dF_t .$$

Moreover, $\varphi_{r,n}\varphi_{s,n} - \varphi_{r+s,n} \to 0$ uniformly on compact subsets of $\mathbb{R}$ as $n \to \infty$. Therefore

$$V_{r+s} = V_r V_s , \quad r, s \geq 0.$$

If $s_n \to s$ then $\varphi_{s_n} \to \varphi_s$ uniformly on compact subsets of $\mathbb{R}$, so

$$V_{s_n} \to V_s \text{ if } s_n \to s , \quad s_n, s \geq 0.$$

Note also that $V_o = I$, $\|V_s\| \leq 1$.

3. Contraction semigroups

A strongly continuous one-parameter semigroup of contractions in $\mathcal{H}$, or simply a contraction semigroup, is a family $\{V_t\}_{t \geq 0} \subset \mathcal{B}(\mathcal{H})$ with

$$\|V_t\| \leq 1 , \quad V_o = I , \quad V_{s+t} = V_s V_t ,$$

and $\quad V_t \to V_s$ if $t \to s$.

We have shown that a maximal dissipative operator gives rise to a contraction semi-group. As for unitary groups and self-adjoint operators, the converse is true.

THEOREM 9.3. There is a 1-1 correspondence between contraction semigroups $\{V_t\}$ and maximal dissipative operators T, given by

$$T = \lim_{\varepsilon \to 0} \frac{1}{i\varepsilon}(V_\varepsilon - I),$$

$$V_s = \int \exp(ist)dF_t \ .$$

Here $\{F_t\}$ is as in Theorem 9.1.

<u>Proof.</u> The procedure is very much as in Chapter 3. Specifically, $D(T) = \{x \mid \lim_{\varepsilon \searrow 0} \frac{1}{i\varepsilon}(T_\varepsilon x - x) \text{ exists}\}$, and Tx is the limit. Exactly as in the proof of Proposition 3.2, T is densely defined. It is easily seen to be dissipative:

$$(Tx, x) = \lim \frac{1}{i\varepsilon}\{(V_\varepsilon x, x) - (x, x)\},$$

so

$$\operatorname{Im}(Tx, x) = \lim \frac{1}{\varepsilon}(\operatorname{Re}[(x, x) - (V_\varepsilon x, x)] \ .$$

But $|(V_\varepsilon x, x)| \leq (x, x)$ since $\|V\ \| \leq 1$, so $\operatorname{Im}(Tx, x) \geq 0$. Again as in the proof of Proposition 3.2 we have

$$(T + i)^{-1} = -i\int_0^\infty e^{-s}V_s \, ds \ .$$

Thus T is maximal dissipative. Now T gives rise to a contraction semigroup $\{V_t'\}$, which gives rise to $T' = \lim_{\varepsilon \to 0} \frac{1}{i\varepsilon}(V_\varepsilon' - I)$. To complete the proof we must show $T = T'$ and $V_t \equiv V_t'$. But

$$(T' + i)^{-1} = -i\int_0^\infty e^{-s}V_s' \, ds = -i\int_0^\infty \int_{-\infty}^\infty \exp(s(it - 1))dF_t \, ds$$

$$= -i\int_{-\infty}^\infty (it - 1)dF_t = \int_{-\infty}^\infty (t + i)^{-1} dF_t$$

$$= (T + i)^{-1} \ ,$$

so $T' = T$. Now if $x \in D(T)$ it is easy to see that $V_t x \in D(T)$, $V_t' x \in D(T)$, and $TV_t x = V_t Tx$, $TV_t' x = V_t' Tx$. It follows that for $x \in D(T)$ the functions $V_t x$ and $V_t' x$ are continuously differentiable. Let $y_t = V_t x - V_t' x$. Then $\frac{d}{dt} y_t = iTV_t x - iTV_t' x = iTy_t$. Therefore

$$\frac{d}{dt}(\|y_t\|^2) = 2\mathrm{Re}(\frac{d}{dt} y_t, y_t) = 2\mathrm{Re}(iTy_t, y_t)$$
$$= -2\,\mathrm{Im}(Ty_t, y_t) \leq 0 .$$

But $y_o = 0$, so $\|y_t\| = 0$. Thus $V_t \equiv V_t'$ on $D(T)$, hence on all of $\mathcal{H}$. ∎

Suppose $\{V_t\}$ is a contraction semigroup in $\mathcal{K}$. A unitary group $\{U_t\}$ in $\mathcal{H} \supset \mathcal{K}$ is said to be a dilation of $\{V_t\}$ if $(V_t x, y) = (U_t x, y)$ for $t \geq 0$, $x, y \in \mathcal{K}$. It is minimal if there is no proper subspace of $\mathcal{H}$ containing $\mathcal{K}$ and reducing each U_t. Theorem 9.2 has an analogue for contraction semigroups. We leave the proof to the reader.

THEOREM 9.4. Any contraction semigroup has a mimimal unitary dilation, which is unique up to unitary equivalence.

10

CHARACTERISTIC FUNCTIONS

1. Characteristic functions of bounded dissipative operators

Suppose $T \in \mathcal{B}(\mathcal{H})$. Then $T = A + iB$ with $A = \operatorname{Re} T$ and $B = \operatorname{Im} T$ self-adjoint. Clearly T is dissipative (and maximal dissipative) if and only if $B \geq 0$. Suppose so. Recall that the linear fractional transformations taking $\mathbb{C}_+$ onto the unit disc and ∞ into 1 are precisely those of the form $\lambda \longmapsto (\lambda - \alpha)(\lambda - \overline{\alpha})^{-1}$, where $\alpha \in \mathbb{C}_+$. In the Cayley transform for operators one takes $\alpha = i$ and lets T play the role of λ. One could instead let T play the role of α and consider

$$\begin{aligned} \lambda \longmapsto (\lambda - T)(\lambda - T^*)^{-1} &= I + (T^* - T)(\lambda - T^*)^{-1} \\ &= I - 2iB(\lambda - T^*)^{-1} \end{aligned}$$

or

$$\lambda \longmapsto (\lambda - T^*)^{-1}(\lambda - T) = I - 2i(\lambda - T^*)^{-1}B \ .$$

Now $2B \geq 0$, so there is a $C \geq 0$ with $C^2 = 2B$. Then we can remove the assymetry here by considering

$$W(\lambda) = I - iC(\lambda - T^*)^{-1}C \quad , \quad \lambda \in \rho(T^*) \ .$$

More generally, it is convenient to consider an auxiliary Hilbert space $\mathfrak{K}$ and $C \in \mathcal{B}(\mathcal{H}, \mathfrak{K})$ with $iC^*C = 2iB = T - T^*$. Given C, set

$$W(\lambda) = I - iC(\lambda - T^*)^{-1}C^* , \quad \lambda \in \rho(T^*) \ .$$

This is called a characteristic function of T. It is said to be reduced if $R(C)$ is dense, i.e. $N(C^*) = (0)$. Since $W(\lambda) \equiv I$ on $N(C^*) = R(C)^\perp$, any characteristic function can be uniquely decomposed as $W_1(\lambda) \oplus I$, where $W_1(\lambda)$ is reduced. We shall call $W_1(\lambda)$ the reduced part of $W(\lambda)$.

THEOREM 10.1. Suppose T is a bounded dissipative operator and $W(\lambda)$ is a characteristic function of T with values in $\mathcal{B}(\mathfrak{k})$. Then

(10.1) $W(\lambda)$ is holomorphic on $\mathbb{C}_+$ and at ∞.

(10.2) $W(\lambda)^* W(\lambda) \leq I$ for $\lambda \in \mathbb{C}_+$.

(10.3) $W(\lambda)^* W(\lambda) = I$ for $|\lambda|$ large, $\lambda \in \mathbb{R}$.

(10.4) There is a $D \geq 0$ in $\mathcal{B}(\mathfrak{k})$ such that $\|W(\lambda) - I + i\lambda D\| = O(|\lambda|^{-2})$ as $|\lambda| \to \infty$.

Conversely suppose $W(\lambda) \in \mathcal{B}(\mathfrak{k})$ is defined for $\lambda \in \mathbb{C}_+$ and for $|\lambda|$ large, and has these properties. Then there is a bounded dissipative operator having $W(\lambda)$ as a characteristic function.

(Strictly speaking, in the converse we assume $W(\lambda)$ defined on a connected set Ω and assert it is the restriction to Ω of a characteristic function of T.)

Proof. Suppose $W(\lambda)$ is a characteristic function of T. Then (10.1) and (10.4) follow from the expression for $W(\lambda)$ and the fact that $(\lambda - T^*)^{-1} = \Sigma \lambda^{-n-1}(T^*)^n$ for $|\lambda| > \|T^*\|$. Also,

$$W(\lambda)^* W(\lambda) = I + iC[(\bar{\lambda} - T)^{-1} - (\lambda - T^*)^{-1}]C^*$$
$$- iC(\bar{\lambda} - T)^{-1}(T - T^*)(\lambda - T^*)^{-1}C$$
$$= I + iC[(\bar{\lambda} - T)^{-1}[\lambda - T^* - \bar{\lambda} + T - T + T^*]\cdot(\lambda - T^*)^{-1}C^*$$
$$= I - 2\,\mathrm{Im}\,\lambda[(\lambda - T^*)^{-1}C^*]^*(\lambda - T^*)^{-1}C^* .$$

This gives (10.2) and (10.3).

Conversely, suppose $W(\lambda)$ has these properties. If $W(\lambda)$ were of the above form for $T = A + iB$, we could consider the related function $V(\lambda) = iC(\lambda - A)^{-1}C^*$. Then

$$W(\lambda) + V(\lambda) = I + iC(\lambda - A)^{-1}(\lambda - T^* - \lambda + A)(\lambda - T^*)^{-1}C$$
$$= I + iC(\lambda - A)^{-1}\,\frac{i}{2}C^*C(\lambda - T^*)^{-1}C$$
$$= I + \frac{1}{2}(V(\lambda) - W(\lambda))$$

or

(10.5) $$V(\lambda) = 2(I - W(\lambda))(I + W(\lambda))^{-1} .$$

We shall show that (10.5) can be used to define $V(\lambda)$, at least on $\mathbb{C}_+$. First we claim that given $\lambda \in \mathbb{C}_+$ there is a $\delta > 0$ such that $\|x + W(\lambda)x\| \geq \delta > 0$ for $\|x\| = 1$. Otherwise there is a sequence $\{x_n\}$ with $\|x_n\| = 1$, $W(\lambda)x_n + x_n \to 0$. Then if $f_n(\mu) = -(W(\mu)x_n, x_n)$, we have f_n holomorphic on $\mathbb{C}_+$, $|f_n(\mu)| \leq 1$, $f_n(\lambda) \to 1$. Therefore $f_n \to 1$ uniformly on compact subsets of $\mathbb{C}_+$. But (10.4) implies $\|I - W(\mu)\| = O(|\mu|^{-1})$ as $\mu \to \infty$, so $f_n(\mu) \to -1$ as $\mu \to \infty$ independent of n. This is a contradiction, and we have that $I + W(\lambda)$ has closed range and, if the range is dense, a bounded inverse. But $R(I + W(\lambda))^{\perp} = N(I + W(\lambda)^*)$. Since $W(\lambda)$ is a contraction, $N(I + W(\lambda)^*) = N(I + W(\lambda))$, and we have shown the latter to be (0). In fact for any contraction W, if $Wx = \pm x$ then

$$\|W^*x \pm x\|^2 = \|x\|^2 + \|W^*x\|^2 \pm 2\mathrm{Re}(W^*x, x)$$
$$= \|x\|^2 + \|W^*x\|^2 + 2\mathrm{Re}(Wx, x)$$
$$\leq \|x\|^2 + \|x\|^2 - 2\|x\|^2 = 0.$$

Thus $V(\lambda)$ is defined by (10. 5) for $\lambda \in \mathbb{C}_+$. It is holomorphic. In fact

$$[I + W(\mu)]^{-1} = (I + W(\lambda))^{-1}[I + (W(\mu) - W(\lambda))(I + W(\lambda))^{-1}]^{-1}$$

is expandable in a convergent power series in $\mu - \lambda$ for $\mu - \lambda$ small, since this implies $\|W(\mu) - W(\lambda)\|$ small. Next note that

$$(10.5)' \qquad \mathrm{Re}\, V(\lambda) = (I + W^*)^{-1}(I - W^*) + (I - W)(I + W)^{-1}$$
$$= (I + W^*)^{-1}[(I - W^*)(I + W) + (I + W^*)(I - W)](I + W)^{-1}$$
$$= 2(I + W^*)^{-1}(I - W^*W)(I + W)^{-1} \geq 0 .$$

Moreover since $W(\lambda) \approx I$ for $|\lambda|$ large and $W^*W = I$ for λ real and $|\lambda|$ large, we have $\mathrm{Re}\, V(\lambda) = 0$ for λ real and $|\lambda|$ large. As $s \to +\infty$,

$$sV(is) = 2s(I - W(is))(I + W(is))^{-1} \to 2D(2I)^{-1} = D .$$

A modification of Theorem 8. 2, with the proof essentially unchanged, shows that there is an operator-valued measure $\{F_t\}$ with $F(\mathbb{R}) = D$ and

$$(10.6) \qquad \mathrm{Re}\, V(\lambda) = \int \mathrm{Im}(t - \lambda)^{-1} dF_t , \qquad \lambda \in \mathbb{C}_+ .$$

Since $F_t \leq F(\mathbb{R}) = D$, all t, this implies that $(\mathrm{Re}\, V(\lambda))x = 0$ if and only if $Dx = 0$. By (10. 5), then, for $\lambda \in \mathbb{C}_+$ we have $\|W(\lambda)x\| = \|x\| \iff Dx = 0$.

We claim that $N(D)$ reduces $W(\lambda)$ for all λ, and $W(\lambda) \equiv I$ on $N(D)$. Take $x \in N(D)$, $y = W(\lambda)x$, $\lambda \in \mathbb{C}_+$. Then $f(\mu) = (W(\mu)x, y)$ is holomorphic on $\mathbb{C}_+$ and $|f(\mu)| \leq \|x\|^2$, $f(\lambda) = \|x\|^2$. Therefore $f(\mu) \equiv \|x\|^2$. Since $\|W(\mu)x\| \equiv \|x\| = \|y\|$, this implies $W(\mu)x \equiv y$. But $W(\mu)x \to x$ as $|\mu| \to \infty$, so $y = x$, $W(\mu)x \equiv x$. By the argument above the fact that $W(\mu)$ is a contraction means $N(D) = N(W(\lambda) - I) = N((W(\lambda) - I)^*)$, $\lambda \in \mathbb{C}_+$. Thus $N(D)$ reduces $W(\lambda)$ for $\lambda \in \mathbb{C}_+$. Let Q be the projection on $N(D)$. Then $0 = W(\lambda)Q - QW(\lambda)$ on $\mathbb{C}_+$, hence on the (connected) domain of definition of $W(\lambda)$, and $W(\lambda) \equiv I$ on $N(D)$.

It follows that we may split off $N(D)$ and assume D to be 1-1. Now $-iV(\lambda)$ and $\int (\lambda - t)^{-1} dF_t$ are holomorphic functions on $\mathbb{C}_+$ with the same imaginary part, and both $\to 0$ as $|\lambda| \to \infty$. Thus

(10.7) $$-iV(\lambda) = \int (\lambda - t)^{-1} dF_t \, , \quad \lambda \in \mathbb{C}_+ .$$

Since $F_t \leq D$, $D^{-1/2} F_t D^{-1/2} \leq I$. Let $F_t' = D^{-1/2} F_t D^{-1/2}$. This is a normalized operator-valued measure. Note also that since $\operatorname{Re} V(\lambda) = 0$ for λ real and large, we have $F_t \equiv 0$ for $t \leq t_o$, $F_t \equiv D$ for $t \geq t_1$. Thus $F_t' \equiv 0$, $t \leq t_o$, $F_t' \equiv I$, $t \geq t_1$. The construction used in Theorem 9.2 produces a Hilbert space $\mathcal{H} \subset \mathfrak{K}$ and a self-adjoint operator A (denoted S previously) such that

(10.8) $$((\lambda - A)^{-1} x, y) = \int (\lambda - t)^{-1} d(F_t' x, y) \, ,$$

$\lambda \in \mathbb{C}_+$, $x, y \in \mathfrak{K}$. Let $P: \mathcal{H} \to \mathfrak{K}$ be the projection, and let $B = \frac{1}{2} DP$, $T = A + iB$. The construction gives $\|A\| \leq \max(|t_o|, |t_1|)$, so T is a bounded dissipative operator.

We claim $W(\lambda)$ is a characteristic operator function for T. In fact let $C = D^{1/2} P$, considered as a mapping from $\mathcal{H}$ to $\mathfrak{K}$. Then

(10.7) and (10.8) show $V_1(\lambda) = V(\lambda)$. Reversing the calculation leading to (10.5) shows that $W_1(\lambda) \equiv W(\lambda)$. ∎

2. Simple dissipative operators and uniqueness

A bounded dissipative operator T acting on $\mathcal{H}$ is said to be <u>simple</u> if there is no non-zero subspace $\mathcal{H}_1 \subset \mathcal{H}$ which reduces T and for which $T|_{\mathcal{H}_1}$ is self-adjoint.

PROPOSITION 10.1. If $T \in \mathcal{B}(\mathcal{H})$ is dissipative, there are unique subspaces $\mathcal{H}_1$, $\mathcal{H}_2$ of $\mathcal{H}$ with $\mathcal{H} = \mathcal{H}_1 \oplus \mathcal{H}_2$, such that $T_1 = T|_{\mathcal{H}_1}$ is simple and $T_2 = T|_{\mathcal{H}_2}$ is self-adjoint.

Proof. Let $\{\mathcal{H}_\alpha\}$ be the collection of all closed subspaces of $\mathcal{H}$ which reduce T and such that the corresponding restriction of T is self-adjoint. Let $\mathcal{H}_2 = \bigvee \mathcal{H}_\alpha$. This also has these properties, and is maximal. It follows easily that the restriction of T to $\mathcal{H}_1 = \mathcal{H}_2^{\perp}$ is simple.

If $\mathcal{H}_1'$ and $\mathcal{H}_2'$ are two other subspaces with the same properties, then by maximality we have $\mathcal{H}_2' \subset \mathcal{H}_2$. By simplicity and the self-adjointness of T on $\mathcal{H}_2$ we get $\mathcal{H}_1' \cap \mathcal{H}_2 = (0)$. Therefore $\mathcal{H}_1' = \mathcal{H}_1$, $\mathcal{H}_2' = \mathcal{H}_2$. ∎

We shall call T_1 and T_2 the <u>simple part</u> and the <u>self-adjoint part</u> of T respectively.

Suppose $W_1(\lambda)$ and $W_2(\lambda)$ are reduced characteristic operator functions for dissipative operators T_1 and T_2, with $W_j(\lambda)$ acting in $\mathcal{K}_2$, j = 2. We shall say that $W_1(\lambda)$ and $W_2(\lambda)$ are <u>equivalent</u> if there is a unitary operator $U: \mathfrak{k}_2 \to \mathfrak{k}_1$ such that $W_2(\lambda) \equiv U^* W_1(\lambda) U$. More generally, if $W_1(\lambda)$ and $W_2(\lambda)$ are not necessarily reduced we shall say they are equivalent if their reduced parts are.

THEOREM 10.2. Let $T_j \in \mathcal{B}(\mathcal{H}_j)$ be dissipative, $j = 1, 2$. Let $W_j(\lambda)$ be a characteristic function of T_j. Then $W_1(\lambda)$ and $W_2(\lambda)$ are equivalent if and only if the simple parts of T_1 and of T_2 are unitarily equivalent.

In particular a simple operator determines an essentially unique reduced characteristic function, and conversely.

Proof. Suppose $W_j(\lambda)$ is determined by $C_j: \mathcal{H}_j \to \mathfrak{k}_j$, $j = 1, 2$. Let T_{11} and T_{12} be the simple and self-adjoint parts of T_1, acting in subspaces $\mathcal{H}_{11}$ and $\mathcal{H}_{12}$ respectively.

Let $T_1 = A_1 + iB_1$, A_1, B_1 self-adjoint. Since T_{12} is self-adjoint, $B_1 \equiv 0$ on $\mathcal{H}_{12}$. But $B_1 = \frac{1}{2} C_1^* C_1$, so $C_1 = 0$ on $\mathcal{H}_{12}$. Therefore $W_1(\lambda)$ coincides with the characteristic function of T_{11} determined by restricting C_1 to $\mathcal{H}_{11}$. It follows that we may assume both T_1 and T_2 are simple. We may also assume that $W_1(\lambda)$ and $W_2(\lambda)$ are reduced.

Suppose $U: \mathcal{H}_2 \to \mathcal{H}_1$ is unitary, $U^* T_1 U = T_2$. Then $U^* B_1 U = B_2$. With $W_1(\lambda)$ and $W_2(\lambda)$ reduced, C_1 and C_2 have dense range. For $x, y \in \mathcal{H}_2$, $(C_2 x, C_2 y) = (C_2^* C_2 x, y) = 2(B_2 x, y)$ $= 2(B_1 Ux, Uy) = (C_1 Ux, C_1 Uy)$. Therefore there is a unique unitary $V: \mathfrak{k}_2 \to \mathfrak{k}_1$ such that $VC_2 = C_1 U$, or $V^* C_1 = C_2 U^*$.

$$\begin{aligned} V^* W_1(\lambda) V &= I - iV^* C_1(\lambda - T_1^*)^{-1} C_1^* V \\ &= I - iC_2 U^*(\lambda - T_1^*)^{-1} U C_2^* \\ &= I - iC_2(\lambda - T_2^*)^{-1} C_2^* \\ &= W_2(\lambda) \ . \end{aligned}$$

Conversely, suppose $W_2(\lambda), W_1(\lambda)$ reduced, T_1, T_2 simple. Suppose there is a unitary $V: \mathcal{K}_2 \to \mathcal{K}_1$ such that $W_2(\lambda) = V^* W_1(\lambda) V$. Let $V_j(\lambda) = iC_j(\lambda - A_j)^{-1} C_j$, as above. Since we can express $V_j(\lambda)$ as a rational function of $W_j(\lambda)$, we get also $V_2(\lambda) = V^* V_1(\lambda) V$. For large $|\lambda|$ the power series expansion and the uniqueness of coefficients gives

$$C_2 A_2^n C_2^* = V^* C_1 A_1^n C_1^* V \ , \quad n = 0, 1, 2, \dots \ . \tag{10.9}$$

Therefore for any polynomials $p(\lambda)$, $q(\lambda)$ and any $x, y \in \mathcal{K}_2$,

$$\begin{aligned} (p(A_2)C_2^* x, q(A_2)C_2^* y) &= (C_2 \bar{q}(A_2) p(A_2) C_2^* x, y) \\ &= (V^* C_1 \bar{q}(A_1) p(A_1) C_1^* V x, y) \\ &= (p(A_1) C_1^* V x, q(A_1) C_1^* V y) \ . \end{aligned} \tag{10.10}$$

Let $\mathcal{H}_j'$ be the smallest closed subspace of $\mathcal{H}_j$ reducing A_j and containing $R(C_j^*)$. By (10.10) there is a unique unitary $U: \mathcal{H}_2' \to \mathcal{H}_1'$ such that $Up(A_2)C_2^* x = p(A_1)C_1^* V x$, $x \in \mathcal{K}_2$, p a polynomial. We shall show that $\mathcal{H}_j' = \mathcal{H}_j$. Assuming this for the moment, applying this to $A_2 p(A_2) C_2^* x$ we find that $UA_2 = A_1 U$, so $A_2 = U^* A_1 U$. Also with $p(\lambda) \equiv 1$ we have $UC_2^* = C_1^* V$. Then $2B_2 = C_2^* C_2 = U^* C_1^* V (U^* C_1^* V)^* = U^* C_1^* C_1 U = 2U^* B_1 U$. Therefore $T_2 = U^* T_1 U$. Finally let us show that $\mathcal{H}_j' = \mathcal{H}_j$. Let $\mathcal{H}_j'' = (\mathcal{H}_j')^{\perp}$. This reduces A_j. Since $R(B_j) \subset R(C_j^*) \subset \mathcal{H}_j'$, $B_j = 0$ on $\mathcal{H}_j''$ and

$\mathcal{H}''_j$ reduces B_j. Therefore $\mathcal{H}''_j$ reduces $T_j = A_j + iB_j$, and $T_j = A_j$ on $\mathcal{H}''_j$. By simplicity, $\mathcal{H}''_j = (0)$. ∎

11

FACTORIZATION AND INVARIANT SUBSPACES

1. The factorization theorem

If $\mathcal{H} = \mathcal{H}_1 \oplus \mathcal{H}_2$ and $S \in \mathcal{B}(\mathcal{H})$, we associate with S the matrix

$$\begin{bmatrix} S_{11} & S_{12} \\ S_{21} & S_{22} \end{bmatrix}$$

where $S_{ij} = P_i S P_j$, P_i the projection on $\mathcal{H}_i$. In particular, if $\mathcal{H}_1$ is <u>invariant</u> for S then $S_{21} = 0$.

Suppose $\mathcal{H} = \mathcal{H}_1 \oplus \mathcal{H}_2$ and $T \in \mathcal{B}(\mathcal{H})$ is dissipative. Suppose also that $\mathcal{H}_1$ is invariant for T. Changing notation slightly we get the associated matrix

$$\begin{bmatrix} T_1 & S \\ 0 & T_2 \end{bmatrix}$$

with $T_j = P_j T P_j$ dissipative also. Associated with T^*, which has $\mathcal{H}_2 = \mathcal{H}_1^{\perp}$ as invariant subspace, is

$$\begin{bmatrix} T_1^* & 0 \\ S^* & T_2^* \end{bmatrix}$$

It is not hard to see that $\rho(T) = \rho(T_1) \cap \rho(T_2)$, and for $\lambda \in \rho(T)$, $(\lambda - T^*)^{-1}$ is given by

$$\begin{bmatrix} (\lambda - T_1^*)^{-1} & 0 \\ (\lambda - T_2^*)^{-1} S^* (\lambda - T_1^*)^{-1} & (\lambda - T_2^*)^{-1} \end{bmatrix}$$

The matrix notation tacitly considers $P_i T P_j$ as an operator from $\mathcal{H}_j$ to $\mathcal{H}_i$. We can and shall also consider it as an operator from $\mathcal{H}$ to $\mathcal{H}$. As such if $T = A + iB$ then clearly $T_j = A_j + iB_j$, where $B_j = P_j B P_j$. Since $P_2 T P_1 = 0$, $S^* = P_2 T^* P_1 = P_2(T^* - T)P_1 = -2P_2 B P_1$. Thus

$$(11.1) \qquad (\lambda - T^*)^{-1} = P_1(\lambda - T_1^*)^{-1} P_1 + P_2(\lambda - T_2^*)^{-1} P_2 - 2iP_2(\lambda - T_2^*)^{-1} P_2 B P_1 (\lambda - T_1^*)^{-1} P_1 .$$

Suppose $C: \mathcal{H} \to \mathfrak{k}$ has $C^* C = 2B$. Let $W(\lambda)$ be the corresponding characteristic function of T. Then $CP_j: \mathcal{H} \to \mathfrak{k}$ has $(CP_j)^* CP_j = 2P_j B P_j = 2B_j$. It follows from this and (11.1) that if $W_j(\lambda)$ is the characteristic function of T_j determined by $C_j = CP_j$, then

$$\begin{aligned} W_2(\lambda) W_1(\lambda) &= (I - CP_2(\lambda - T_2^*)^{-1} P_2 C^*)(I - iCP_1(\lambda - T_1^*)^{-1} P_1 C^*) \\ &= \ldots \\ &= W(\lambda) . \end{aligned}$$

Note that since $T_j = P_j T P_j = 0$ on $\mathcal{H}_j^{\perp}$, the simple part of T_j is the same as the simple part of the restriction of $P_j T$ to $\mathcal{H}_j$. Thus we have the first half of the factorization theorem for characteristic functions.

THEOREM 11.1. Suppose $T \in \mathcal{B}(\mathcal{H})$ is dissipative. Suppose $\mathcal{H}_1$ and $\mathcal{H}_2$ are complementary subspaces of $\mathcal{H}$ and that $P_j: \mathcal{H} \to \mathcal{H}_j$ is the projection. If $\mathcal{H}_1$ is invariant for T, then any characteristic function of T has a factorization $W(\lambda) = W_2(\lambda)W_1(\lambda)$, where $W_j(\lambda)$ is a characteristic function for $T_j = P_jT|_{\mathcal{H}_j}$.

Conversely, suppose $T_j \in \mathcal{B}(\mathcal{H}_j)$ is dissipative, $j = 1, 2$, and suppose $W_j(\lambda)$ is a characteristic function of T_j acting in $\mathfrak{k}$, $j=1, 2$. Then there is an operator $T \in \mathcal{B}(\mathcal{H}_1 \oplus \mathcal{H}_2)$ such that $W_2(\lambda)W_1(\lambda)$ is a characteristic function of T. Moreover $\mathcal{H}_1$ is invariant for T and $T_j = P_jT|_{\mathcal{H}_j}$ where $P_j: \mathcal{H} \to \mathcal{H}_j$.

Proof. Suppose $W_j(\lambda)$ acting in $\mathfrak{k}$ is a characteristic function for $T_j \in \mathcal{B}(\mathcal{H}_j)$. Let $\mathcal{H} = \mathcal{H}_1 \oplus \mathcal{H}_2$. We want to take T corresponding to the matrix

$$\begin{bmatrix} T_1 & S \\ 0 & T_2 \end{bmatrix},$$

where $S \in \mathcal{B}(\mathcal{H}_2, \mathcal{H}_1)$ is to be determined. With $T_j = A_j + iB_j$, we would have $T = A + iB$, B given by

$$\begin{bmatrix} B_1 & \frac{1}{2i}S \\ -\frac{1}{2i}S^* & B_2 \end{bmatrix}.$$

Moreover we want $C: \mathcal{H} \to \mathfrak{k}$ with $CP_j = C_j$, $C^*C = 2B$. But $\frac{1}{2}C^*C$ has matrix

$$\begin{bmatrix} C_1^*C_1 & C_1^*C_2 \\ C_2^*C_1 & C_2^*C_2 \end{bmatrix} = \begin{bmatrix} B_1 & \frac{1}{2}C_1^*C_2 \\ \frac{1}{2}(C_1^*C_2)^* & B_2 \end{bmatrix}.$$

Thus we can take $S = iC_1^* C_2$. The same direct computation as before shows that if $W(\lambda)$ is the characteristic function for T determined by C, then $W(\lambda) = W_2(\lambda)W_1(\lambda)$. Moreover $\mathcal{H}_1$ is invariant for T and $T_j = P_j T|_{\mathcal{H}_j}$. ∎

Note that the theorem does not assert that a factorization of a characteristic function for T leads to an invariant subspace for T; it leads to an invariant subspace for an operator which has the same characteristic function. If both T and the constructed operator are simple, then they are unitarily equivalent and we do get an invariant subspace for T.

Note also that by Theorem 10.1 we can say when a given factorization $W(\lambda) = W_2(\lambda)W_1(\lambda)$ is a factorization into characteristic functions of dissipative operators.

2. Compact dissipative operators

Recall that a bounded operator $T: \mathcal{H} \to \mathcal{K}$ is said to be compact (or completely continuous) if it takes bounded sets to sets with compact closure. Linear combinations and adjoints of compact operators are compact. If $T \in \mathcal{B}(\mathcal{H}_1, \mathcal{H}_2)$ is compact and $S \in \mathcal{B}(\mathcal{H}_2, \mathcal{H}_3)$ (or $S \in \mathcal{B}(\mathcal{H}_3, \mathcal{H}_1)$) then ST (or TS) is compact. Also T is compact if and only if T^*T is compact. We need one more fact: a compact self-adjoint operator has an eigenvalue.

LEMMA 1. A compact dissipative operator T is simple if and only if it has no real eigenvalues.

Proof. Suppose T is not simple. Then its self-adjoint part is compact and non-trivial, hence has a real eigenvalue.

Conversely, suppose $Tx = \lambda x$ for some $x \neq 0$ and $\lambda \in \mathbb{R}$. Then $(Bx, x) = \mathrm{Im}(Tx, x) = 0$, so $B^{1/2}x = 0$, so $Bx = 0$. Then $T^{*}x = (T - 2iB)x = \lambda x$. Therefore the subspace spanned by x reduces T and the restriction of T to this subspace is self-adjoint. ∎

LEMMA 2. Suppose $T \in \mathcal{B}(\mathcal{H})$ is compact and dissipative. Suppose $\mathcal{H} = \mathcal{H}_1 \oplus \mathcal{H}_2$, $P_j: \mathcal{H} \to \mathcal{H}_j$ the projection, and $T_j = P_j T$ restricted to $\mathcal{H}_j$. If $\mathcal{H}_1$ is invariant for T, then T is simple if and only if both T_1 and T_2 are.

Proof. If T_1 is not simple, then it has a real eigenvalue. But $T|_{\mathcal{H}_1} = T_1$, so T does also. If T_2 is not simple, then similarly T^* has a real eigenvalue. Then the dissipative operator $-T^*$ is not simple, so T is not simple.

Conversely, suppose T is not simple. Then there is $\lambda \in \mathbb{R}$, $x \neq 0$ with $Tx = \lambda x$. Write $x = x_1 + x_2$ with $x_j \in \mathcal{H}_j$. Then $\lambda x_1 + \lambda x_2 = Tx = (T_1x_1 + P_1Tx_2) + T_2x_2$. The term in parenthesis is in $\mathcal{H}_1$, so $\lambda x_2 = T_2x_2$. If $x_2 \neq 0$, then T_2 is not simple. If $x_2 = 0$, then $\lambda x_1 = Tx_1$ and T_1 is not simple. ∎

For simple compact operators we can complete the discussion of the relationship between invariant subspaces and factorizations. Let us say that two subspaces $\mathcal{H}_1$ and $\mathcal{H}_1'$ which are invariant for the dissipative operator $T \in \mathcal{B}(\mathcal{H})$ are <u>conjugate</u> if there is a unitary $U \in \mathcal{B}(\mathcal{H})$ such that $UT = TU$ and $U(\mathcal{H}_1') = \mathcal{H}_1$.

We shall say that two factorizations of a characteristic function, $W(\lambda) = W_2(\lambda)W_1(\lambda)$ and $W(\lambda) = W_2'(\lambda)W_1'(\lambda)$ are <u>equivalent</u> if there is a unitary $V \in \mathcal{B}(\mathfrak{K})$ such that $W_j'(\lambda) = V^*W_j(\lambda)V$, $j = 1, 2$.

THEOREM 11.2. Let $T \in \mathcal{B}(\mathcal{H})$ be a compact simple dissipative operator with characteristic operator function $W(\lambda)$. Two invariant subspaces for T give rise to equivalent factorizations of $W(\lambda)$ if and only if they are conjugate.

<u>Proof.</u> Let $\mathcal{H}_1$ and $\mathcal{H}'_2$ be invariant for T, with orthogonal complements $\mathcal{H}_2$ and $\mathcal{H}'_2$. Let P_j, P_j' be the corresponding projection, and suppose $W(\lambda)$ is determined by $C: \mathcal{H} \to \mathfrak{K}$. Suppose first that $\mathcal{H}_1$ and $\mathcal{H}'_1$ are conjugate. Take U unitary so that $UT = TU$ and $U(\mathcal{H}'_1) = \mathcal{H}_1$. Now the identical operators T and U^*TU have the same characteristic function determined by C, so the proof of Theorem 10.2 shows that there is a unitary V in $\mathfrak{K}$ such that $V^*C = CU^*$. Since $U(\mathcal{H}'_1) = \mathcal{H}_1$, we have $UP_j' = P_jU$. Now

$$W_j(\lambda) = I - iCP_j(\lambda - T^*)^{-1}P_jC^*.$$

Multiply by V^* on the left and by V on the right. We have $V^*CP_j = CU^*P_j = CP_j'U^*$, $P_jC^*V = P_jUC^* = UP_j'C^*$, and $U^*(\lambda - T^*)^{-1}U = (\lambda - T^*)^{-1}$. Thus $V^*W_j(\lambda)V = W_j'(\lambda)$.

Conversely, suppose the factorizations induced by $\mathcal{H}_1$ and $\mathcal{H}'_1$ are equivalent. Since T_j and T_j' are simple, by Lemma 2, there are unique unitary operators $U_j: \mathcal{H}'_j \to \mathcal{H}_j$ such that $T_jU_j = T_j'U_j'$ and such that

$$U_j(C_j')^* = C_j^*V \quad , \quad C_jU_j = VC_j' \quad ,$$

where V gives the equivalence of the factorization and $C_j = CP_j$, $C_j' = CP_j$. Identify $\mathcal{H}$ with $\mathcal{H}_1 \oplus \mathcal{H}_2$ and with $\mathcal{H}_1' \oplus \mathcal{H}'$ the usual way, and let $U = U_1 \oplus U_2$. As in the proof of Theorem 11.1 $P_1TP_2 = iC_1^*C_2$. Then

$$\begin{aligned} TU &= (P_1TP_1 + P_1TP_2 + P_2TP_2)U \\ &= T_1U_1P_j' + iC_1^*C_2U_2P_2' + T_2U \\ &= U_1T_1'P_1' + iU_1(C_1')^*C_2'P_2 + U_2T_2'P_2 \\ &= UT, \end{aligned}$$

while $U(\mathcal{H}_1') = U_1(\mathcal{H}_1') = \mathcal{H}_1$. Thus $\mathcal{H}_1$ and $\mathcal{H}_1'$ are conjugate. ∎

Recall that the rank of an operator $S \in \mathcal{B}(\mathcal{H})$ is the dimension of its range.

COROLLARY. Let $T, \mathcal{H}_1, \mathcal{H}_1', W(\lambda)$ be as in Theorem 11.1. Suppose $\operatorname{Im} T$ has rank 1. Then $\mathcal{H}_1$ and $\mathcal{H}_1'$ lead to equivalent factorizations of $W(\lambda)$ if and only if they coincide.

Proof. It suffices to show that if U is unitary and commutes with T, then U is a scalar multiple of I. But if $UT = TU$, then $UT^* = U(UT)^*U = U(TU)^*)U = T^*U$. Therefore U commutes with $B = \operatorname{Im} T$. Let $\mathcal{K} = B(\mathcal{H})$. This is one-dimensional and an eigenspace for B, hence invariant for U. Multiplying by a scalar, we may assume $Ux = x$ for $x \in \mathcal{K}$. Now also U commutes with $A = \operatorname{Re} T$. Then for $x \in \mathcal{K}$, $UA^n x = A^n Ux = A^n x$. Since T is simple, the $A^n x$, $x \in \mathcal{K}$, span a dense subspace of $\mathcal{H}$. Therefore $U = I$. ∎

3. An application: a Volterra integral operator

Let $\mathcal{H} = L^2([0,1])$ and let $T \in \mathcal{B}(\mathcal{H})$ be the operator with

$$Tf(t) = i\int_0^t f(s)ds. \tag{11.2}$$

Then

$$T^*f(t) = -i\int_t^1 f(s)ds,$$

so if $B = \operatorname{Im} T$, $Bf(t) = \frac{1}{2}\int_0^1 f(s)ds = \frac{1}{2}(f, e)e(t)$, where $e(t) \equiv 1$.

Thus B has rank 1. We claim that T is compact. In fact by the Schwarz inequality,

$$|Tf(t') - Tf(t)| = |\int_{t'}^t f(s)ds| \le \|f\|(t - t')^{1/2}.$$

Thus T takes bounded sets in $\mathcal{H}$ to bounded, uniformly equicontinuous sets of continuous functions. By the Ascoli-Arzela theorem the latter are pre-compact in $C([0,1])$, hence pre-compact in $\mathcal{H}$.

The subspaces

$$\mathcal{H}_a = \{f \in \mathcal{H} \mid f = 0 \text{ a.e. on } [0,a]\}, \quad 0 \le a \le 1, \tag{11.3}$$

are clearly invariant for T.

THEOREM 11.3. Every closed subspace of $\mathcal{H}$ invariant for the operator T given by (11.2) is of the form (11.3).

Proof. By induction

$$T^n f(t) = i^n \int_0^t \frac{(t-s)^{n-1}}{(n-1)!} f(s)ds, \quad n = 1, 2, \ldots .$$

Then for $|\lambda|$ large,

$$(\lambda - T)^{-1}f(t) = \sum_{n=0}^{\infty} \lambda^{-n-1}T^n f(t)$$
$$= \lambda^{-1}f(t) + i\lambda^{-2}\int_0^t \sum_{n=0}^{\infty} \frac{1}{n!}\left[\frac{i(t-s)}{\lambda}\right]^n f(s)ds$$
$$= \lambda^{-1}f(t) + i\lambda^{-2}\int_0^t \exp\left(\frac{i(t-s)}{\lambda}\right)f(s)ds .$$

Let $\mathfrak{k}$ be the subspace of $\mathcal{H}$ spanned by e, and let $C: \mathcal{H} \to \mathfrak{k}$ be given by $f \mapsto (f, e)e$. Then $C^*C = 2B = 2\,\mathrm{Im}\,T$. The functions in $\mathfrak{k}$ are constant, and we shall identify $\mathfrak{k}$ with $\mathbb{C}$ in the natural way. Then $\mathcal{B}(\mathfrak{k})$ is identified with $\mathbb{C}$, also. Thus the characteristic function determined by C is a scalar function $w(\lambda)$. A calculation shows

$$(\lambda - T)^{-1}e(t) = \lambda^{-1}\exp(i\lambda^{-1}t), \qquad \lambda \neq 0.$$

Since these functions span a dense subspace of $\mathcal{H}$, T is simple. We have

$$w(\lambda) = ([i - iC(\lambda - T^*)^{-1}C^*]e, e)$$
$$= 1 - i((\lambda - T^*)^{-1}e, e) = 1 - i(e, (\bar{\lambda} - T)^{-1}e)$$
$$= 1 - i\lambda^{-1}\int_0^1 \exp(i\lambda^{-1}t)dt = \exp(i\lambda^{-1}).$$

An analogous computation for the restriction T_a of T to $\mathcal{H}_a$ shows that it has a scalar characteristic function

(11.4) $$w_a(\lambda) = \exp(i(1-a)\lambda^{-1}).$$

To complete the proof we need only show that if $\mathcal{H}'$ is invariant for T and T' is the restriction, then the scalar characteristic function $w'(\lambda)$ is of the form (11.4) for some $a \in [0, 1]$. Now $w'(\lambda)$ is holomorphic on $\mathbb{C}_+$ and divides $w(\lambda)$, hence has no zeros.

Moreover $|w'(\lambda)| \leq 1$ on $\mathbb{C}_+$. Therefore $g(\lambda) = -\log|w(\lambda)|$ $= -\mathrm{Re}\log w'(\lambda)$ is harmonic and non-negative on $\mathbb{C}_+$. Since T' is compact and simple, every non-zero real λ is in $\rho(T')$, hence in $\rho((T')^*)$. Therefore $w'(\lambda)$ is holomorphic on $\mathbb{R} - \{0\}$. Since $(w'(\lambda))^{-1}w(\lambda)$ is also a characteristic function, $1 \geq |w'(\lambda)| \geq |w(\lambda)| = 1$, $\lambda \in \mathbb{R} - \{0\}$. Thus $g(\lambda) = 0$ on $\mathbb{R}$ except possibly at $\{0\}$. It follows (see Chapter 8) that g is of the form $b\,\mathrm{Im}\,\lambda - c\,\mathrm{Im}\,(\lambda^{-1})$, where $b, c > 0$. Since $w'(\lambda) \to 1$ as $|\lambda| \to \infty$, $g(\lambda) \to 0$ as $|\lambda| \to \infty$, so $b = 0$. Thus $\log w'(\lambda) = ic\lambda^{-1} + id$, $d \in \mathbb{R}$. But if we choose the branch of log with $\log 1 = 0$, we get $\log w'(\lambda) \to 0$ as $|\lambda| \to \infty$, so $d = 0$. Thus $w'(\lambda) = \exp(ic\lambda^{-1})$. Finally, $|w'(\lambda)| \leq 1$, all $\lambda \in \mathbb{C}_+$ implies $0 \leq c \leq 1$. Taking $a = 1 - c$, we get $w'(\lambda) = w_a(\lambda)$, so $\mathcal{H}' = \mathcal{H}_a$. ∎

An operator whose invariant subspaces are totally ordered with respect to inclusion is said to be <u>unicellular</u>. We have shown that the Volterra integral operator (11.2) is unicellular.

NOTES AND BIBLIOGRAPHY

Notes

Chapter 1. Standard references for the theory of linear operators in Hilbert spaces are the books of Ahiezer and Glazman [1], Dunford and Schwartz [8], Riesz and Sz.-Nagy [17], and Stone [19].

Chapters 2 and 3. The functional calculus and spectral theorem for unbounded self-adjoint operators are treated in [1], [7], [17], and [19]. The derivation given here, based directly on the algebra $\mathfrak{F}$ of rational functions, may be new in some details. (There are so many proofs of the spectral theorem that this is, a priori, unlikely.) A second approach to these topics is given in chapters 8 and 9.

Chapter 4. The L^2-space of vector-valued functions is a special case of the Bochner integral; see chapter 3 of the book of Hille and Phillips [13]. Proposition 4.3 is a generalization of a theorem of Fuglede. Various Hilbert space versions of Propositions 4.4 and 4.5 seem to be folk theorems. See [2] for a more general setting of Propositions 4.3, 4.4, and 4.5 and related results.

Chapter 5. The treatment here of spectral representation and spectral multiplicity is similar to that in chapter 12 of [8]; see also [19]. For a quite different treatment, see [10].

Chapter 6. For a general discussion of discrete and continuous shifts, see also the paper of Halmos [11]. The application to the F. and M. Riesz theorem was suggested by the paper of Forelli [9], which contains an interesting generalization. Much more information on invariant subspaces is contained in the book of Helson [12]. The original (multiplicity one) theorem is in Beurling [3]. For this and the structure of inner functions, see also the book of Hoffman [14].

Chapter 7. Theorem 7.1 is due to Sinai, who reduced it to the theorem of von Neumann. The reduction to the theorem for the (discrete) shift is due to Lax and Phillips; see [15]. von Neumann's theorem is a special case of Mackey's theorem on induced representations. For a treatment from this point of view, and the application to a general Plancherel theorem, see the survey article of Segal [18]. The invariant subspace theorem for continuous shifts was first stated by Lax; see [11] and [12].

Chapter 8. This is mainly folk theory. Some of the ideas are already evident in Stone's book [19].

Chapter 9. The treatment of a dissipative operator typically uses the Cayley transform to reduce it to a contraction. The present treatment is intended to show that this is neither natural nor necessary. Theorem 9.1 is given in Dolph and Penzlin [6], but again the ideas go back at least to Stone's treatment of the spectral theorem. Theorem 9.2 is the counterpart of Sz.-Nagy's theorem on unitary dilations of contractions. For the latter and a development of the functional calculus from that point of view see the book of Sz.-Nagy and Foiaş [16].

For a still different viewpoint, see the paper of de Branges [4]. Theorem 9.3 is the Hilbert space version of the Hille-Yosida theorem; see [13].

Chapters 10 and 11. The "characteristic function" arose in the theory of differential operators. The abstract version is due to Livšic. A definition applicable to unbounded operators is given in the paper of Štraus [20]. Essentially the same function arises in the theory of contractions; see [15]. A general reference for chapters 10 and 11 is the survey article of Brodskii and Livšic [5], to whom most of these results are due. The treatment of the Volterra integral operator here is rather different from that in [5].

Bibliography

1. Ahiezer, N. I., and Glazman, I. M. Theorie der linearen Operatoren im Hilbert-Raum. Berlin: Akademie-Verlag, 1954.

2. Beals, R. Operators in function spaces which commute with multiplications. Duke Math. J. 35 (1968): 353-62.

3. Beurling, A. On two problems concerning linear transformations in Hilbert space. Acta Math. 81 (1949): 239-55.

4. Branges, L. de. Factorization and invariant subspaces. J. Math. Anal. Appl. 29 (1970): 163-200.

5. Brodskii, M. S., and Livšic, M. S. Spectral analysis of non-self-adjoint operators and intermediate systems. Uspehi Mat. Nauk 13 (1958): 1-85; Amer. Math. Soc. Translations, ser. 2, vol. 13 (1960): 265-346.

6. Dolph, C. L., and Penzlin, F. On the theory of a class of non-self-adjoint operators and its applications to quantum scattering theory. Ann. Acad. Sci. Fennicae, No. 263 (1959).

7. Douglas, R. G. Structure theory for operators, I. J. Reine Angew. Math. 232 (1968): 180-93.

8. Dunford, N., and Schwartz, J. T. Linear operators, part 2, New York: Interscience, 1963.

9. Forelli, F. Analytic and quasi-invariant measures. Acta Math. 118 (1967):33-59.

10. Halmos, P. Introduction to Hilbert space and the theory of spectral multiplicity. New York: Chelsea, 1951.

11. Halmos, P. Shifts on Hilbert space. J. Reine Angew. Math. 208 (1961):102-12.

12. Helson, H. Lectures on invariant subspaces. New York: Academic Press, 1964.

13. Hille, E., and Phillips, R.S. Functional analysis and semi-groups. Colloquium Publications, vol. 31. Providence, R.I.: American Mathematical Society, 1957.

14. Hoffman, K. Banach spaces of analytic functions. Englewood Cliffs, N.J.: Prentice Hall, 1962.

15. Lax, P.D., and Phillips, R.S. Scattering theory. New York: Academic Press, 1967.

16. Sz.-Nagy, B., and Foiaş, C. Analyse harmonique des opérateurs dans l'espace de Hilbert. Paris: Masson, 1967.

17. Riesz, F., and Sz.-Nagy, B. Functional analysis. New York: Ungar, 1955.

18. Segal, I.E. Algebraic integration theory. Bull. Amer. Math. Soc. 71(1965): 419-89.

19. Stone, M. H. Linear transformations in Hilbert space and their applications to analysis. Colloquium Publications, vol. 15. New York: American Mathematical Society, 1932.

20. Strauss, A. V. Characteristic functions of linear operators. Izv. Akad. Nauk SSSR, ser. mat. 24 (1960): 43-74; Amer. Math. Soc. Translations, ser. 2, vol. 40 (1964): 1-37.

NOTATION

INDEX